Cultural Industries and the Environmental Crisis

Kate Oakley • Mark Banks
Editors

Cultural Industries and the Environmental Crisis

New Approaches for Policy

Springer

Editors
Kate Oakley
University of Glasgow
Glasgow, UK

Mark Banks
University of Glasgow
Glasgow, UK

ISBN 978-3-030-49383-7 ISBN 978-3-030-49384-4 (eBook)
https://doi.org/10.1007/978-3-030-49384-4

This Springer imprint is published by the registered company Springer Nature Switzerland AG
The registered company address is: Gewerbestrasse 11, 6330 Cham, Switzerland

Acknowledgements

Kate would like to thanks colleagues on the CUSP programme, particularly Tim Jackson, Ian Christie, Mark Ball, Malaika Cunningham and Jonathan Ward.

Mark would like to thank former colleagues at the CAMEo Research Institute for Cultural and Media Economies at the University of Leicester, especially Isaac Hoff for his research assistance in preparing this book.

Contents

The original version of this book was revised. The correction is available at https://doi.org/10.1007/978-3-030-49384-4_12

List of Contributors

Mark Banks is Professor of Cultural Economy at the University of Glasgow where he researches cultural industries, work and employment, creative economy, and cultural policy. He is the author of Creative Justice: Cultural Industries, Work and Inequality (2017) and The Politics of Cultural Work (2007).

Matt Brennan is Reader in Popular Music at the University of Glasgow. He publishes widely on popular music and society and is the author of two monographs, *Kick It: A Social History of the Drum Kit* (2020) and *When Genres Collide* (2017).

Brett Caraway is a Professor of Information and Telecommunications Economics at the University of Toronto in the Institute of Communication, Culture, Information and Technology. His research examines recurrent crises with respect to the economic, social, and ecological impacts of information and communication technology.

Cecilia Dinardi is a Senior Lecturer in Cultural Policy and Arts Administration at ICCE, Goldsmiths, University of London. Her areas of research include culture led urban regeneration, the informal creative economy, particularly in Latin America, and ethnographic documentary filmmaking.

Martina Drosner is a professional interested in creative and resilient cities. Between 2017 and 2020, Martina worked as a Project Manager at Biovision Foundation for Ecological Development and as Communication Manager at the Sustainable Development Solutions Network (SDSN) Switzerland in Zurich.

Jonathan Gross is a Research Fellow in the Department of Culture, Media and Creative Industries, King's College London. His work addresses questions of cultural policy, politics, and participation.

Anthony Killick is a Lecturer in Media, Culture and Communication at Liverpool John Moores University.

Mariangela Lavanga is a Senior Assistant Professor of Cultural Economics at the Erasmus University Rotterdam. She is Academic Co-ordinator of the Master Cultural Economics and Entrepreneurship and Co-founder and Co-ordinator of the Minor Fashion Industry.

Richard Maxwell is Professor of Media Studies at Queens College, City University of New York. His publications include The Spectacle of Democracy, Culture Works: The Political Economy of Culture, Herbert Schiller, Global Hollywood (co-author), Greening the Media (with Toby Miller), The Routledge Companion to Labor and Media, Media and the Ecological Crisis (co-editor), and How Green is Your Smartphone? (with Toby Miller).

Ealasaid Munro Her research focuses primarily on the experience of working within media and cultural industries. She currently works outside academia.

Kate Oakley is Professor of Cultural Policy and Head of the School of Culture and Creative Arts at the University of Glasgow.

Frederick Harry Pitts is a Lecturer in Work, Employment, Organisation and Public Policy at the University of Bristol School of Management, where he also leads the Faculty of Social Sciences and Law Research Group for Perspectives on Work and co-edits the Bristol University Press online magazine Futures of Work. He is the author of Value (Polity, 2020), Corbynism: A Critical Approach (Emerald, 2018), and Critiquing Capitalism Today (Palgrave, 2017).

Paula Serafini is Research Associate at CAMEo Research Institute for Cultural and Media Economies, University of Leicester, and her research is in the field of cultural politics. She is the author of Performance Action: The Politics of Art Activism (Routledge 2018).

Nick Wilson is Head of Department, Culture, Media and Creative Industries (CMCI) and Professor of Culture and Creativity at King's College London.

Chapter 1
Cultural Industries and Environmental Crisis: An Introduction

Kate Oakley and Mark Banks

1.1 Introduction

Culture and the arts—where they are considered at all in environmental debates—are generally viewed as either benign low carbon activities that bring pleasure and meaning, or as irrelevant in the face of existential crisis. As cultural industry scholars, we reject both these readings and instead argue for a critical consideration of the role and potential of cultural activities in the face of mounting crises, environmental and otherwise. At the very least, cultural industries are part of the way we make sense of things and sense making is as vital as ever, but in addition they are huge commercial entities, instruments of public policy across the globe, and, in some cases, major polluters and resource consumers.

This edited collection will explore these issues—at the nexus of cultural and environmental concerns—from a variety of angles. Its origins are in a five-year interdisciplinary research project, the Centre for Understanding Sustainable Prosperity (CUSP), based in the UK but drawing on an international network of scholars. CUSP's concern is how we might live well and even flourish on planet of resource constraints and amidst mounting social problems. The project has explored this concern in various ways, including through studies of sustainable finance, 'good work' and the moral underpinnings of environmentalism. One of us (Oakley) has led a team examining the role of arts and culture in advancing what CUSP Director Tim Jackson (2016) calls 'sustainable prosperity'. In this, we have explored ways in which arts and cultural activities can help develop ideas of the 'good life' beyond

K. Oakley (✉) · M. Banks
University of Glasgow, Glasgow, UK
e-mail: kate.oakley@glasgow.ac.uk; mark.banks@glasgow.ac.uk

K. Oakley, M. Banks (eds.), *Cultural Industries and the Environmental Crisis*,
https://doi.org/10.1007/978-3-030-49384-4_1

unsustainable material consumption and researched the complex interactions between cultural prosperity, place and the quality (and availability) of employment and leisure. Throughout the project we have tried to argue that only by addressing deep-rooted social problems—inequality in particular—can we hope to survive the environmental crisis and that only by holding on to a vision of a better future can such an enormous effort be mobilized. This normative project is very much at the heart of this edited collection as well.

This introductory chapter provides some context for the book and looks in particular at the orthodox discourse of the 'creative economy' and the role it has played in establishing what we argue is an unsustainable model of cultural industries production and consumption. The final section of the introduction will look in more detail at the individual chapters and some of the shared themes they intimate or suggest. In keeping with the overall CUSP project (and our own intellectual ambitions) we strive also to explore some avenues for hope, identifying instances where environmental (and wider ecological) crisis might provide opportunities as well as threats, and where cultural producers are helping to progressively address the urgent dilemmas and challenges now being raised by ecological change under advanced capitalism.

1.2 Entwined Crises

We are well into the second decade of a global financial crisis and what was initially dismissed as a short-term 'credit crunch', and then a temporary 'downturn', now seems more like a terminal stasis or decline, sometimes referred to as 'secular stagnation' (Jackson 2018). GDP per capita and labour productivity across the richest nations have been declining for decades and the Global North look set for a period of sustained low growth rates, just at a time when mounting social problems require large-scale public spending and investment. However, governments and mainstream economists appear unwilling to accept the need for any reappraisal of dominant ideas, even as nations continue to suffer the effects of multiple and repeating economic crises that appear increasingly damaging, yet also immovable (Jacobs and Mazzucato 2016).

Together, and relatedly, the durability of some ingrained and seemingly unchallengeable inequalities has precipitated social crisis. While overall rates of growth might plateau or decline, the very richest minority has become more adept at securing a disproportionate share of the available wealth, so entrenching poverty, low wages, and a reduction of welfare supports for the poorer majority (Piketty 2014). Material deprivations, coupled with some skillful political manipulation of public anxieties, has fueled the rise of right and far right governments in a number of countries, and raised wider concerns about the stability of social democracy and any future provision of social systems that might at least aspire to provide all citizens with shared opportunity and a collective well-being.

Furthermore, the last years of the 2010s saw a huge increase in media reporting of some of the environmental problems being occasioned by a capitalist-fuelled climate crisis, propelled in part by mounting number of environmental disasters themselves—Australian bushfires, Indonesian floods at time of writing—but also by growing, media-savvy resistance in the form of the school strikes led by Greta Thunberg. This volume is not concerned with media representation of the climate crisis *per se*. Rather, the point here is that while the climate crisis is, and has been, intensifying for some time, *public awareness* of the crisis is probably at an all-time high. The sense of crisis is a pervasive one and is reflected increasingly in the texts and commentaries of media and cultural producers, as well as in the aesthetic politics of emerging organizations like Extinction Rebellion. Yet, at the same, time the mainstream media are also accused of exacerbating crisis, through the pervasive reduction of environmental disasters to banal 'images of devastation' that encourage a 'distanced voyeurism' (Demos 2018) which displaces discussions of responsibility, and complicity, and the significant political steps that will be necessary to reverse the accelerating pace of destruction.

It therefore seems a perfect time for the cultural industries to be thinking not only about what to say about these entwined crises, but also about developing different and sustainable models of *sourcing, organizing, producing and distributing* their cultural content in order to help challenge or alleviate crisis. For us, this assumes a need to prioritize ways of making cultural goods that don't rest on assumptions of unchecked accumulation or economic growth, nor the unlimited extraction or exploitation of physical and human resources, but seek instead to challenge these conventional understandings, norms and practices. But this is generally not happening. While there are now an encouraging number of 'socially-just', 'inclusive', 'ecological' 'transitional', 'decolonized' and 'post-growth' approaches to cultural industries production, these still remain fairly small-scale, marginal and disintegrated. At the same time, the pursuit of 'business as usual' has produced a situation where the cultural and media industries are now major polluters, and where digital technologies have overtaken aerospace in the production of greenhouse gases and video streaming services are the emissions equivalent of Spain.

In capitalist economies there is nothing new about cultural production being at the forefront of new technology use, nor about their promotion of an idea of growth and expansion. But alongside that, the belief in the arts and culture as source of resistance or alternatives remains resilient, and while the counter culture of the 1960s may have collapsed into Silicon Valley tech-dystopias, plenty of artists, writers and musicians work continually to ensure that the sense of culture as a space of progressive possibility is not entirely extinguished, and that their own work is able to draw attention to the damages and dangers of unchecked capitalist cultural production. In parallel, within academia, environmental concerns, particularly about energy use, the extractive intensity of new digital technologies, and electronic waste (Gabrys 2011; Maxwell and Miller 2012) have become more prominent and the complicity of the cultural industries with unsustainable models of economic growth is now being questioned. We are witnessing the rise of new approaches that seek to challenge the idea of economic growth as a desirable social objective, drawing on recently revived notions of 'degrowth' (for example, see Kallis 2018) or otherwise prioritizing cultural economies that foreground commitments to overall human well-being rather than GDP and accumulation (Banks 2018; Oakley et al. 2018). And so while critics of the cultural industries might once have had relatively little to say about such matters, recent years have seen a new critical consciousness emerge that has sought to engage with the social, environmental and ecological limits of creative, media and cultural industry growth (Caraway 2017; Maxwell and Miller 2017; Oakley and Ward 2018). This is taking on a particular urgency and resonance, not least because governments and creative economy policy-makers ceaselessly tend to promote creative and cultural industries as the best guarantee of ensuring the much desired return to an unshackled (and therefore unsustainable) pre-crisis economy (for example, see Bazalgette 2017; DDCMS 2017).

1.3 Cultural Industries and Creative Economy

While, for most commentators, global capitalism remains in the throes of crisis, this is far from detectable in orthodox policy discourse on the cultural or creative sectors. Just in the UK alone, the current UK Industrial Strategy, the Creative Industries Sector Deal, national Governmental research programmes, the Creative Industries Federation, and the British Council abroad, remain committed to talking-up the creative economy, routinely identifying it as the 'fastest growing sector' and at the forefront of (a much anticipated, but never quite arriving) national economic recovery. Much the same views can be found in other national policy contexts (for example, see BMWi 2017; Canada Heritage 2017; France Créative 2013) and transnational bodies such as UNESCO and UNCTAD (2018) and the EU (2018) continue to promote culture and creativity as commercial panaceas.

While even in the most stagnant national economies the creative economy is reported as booming, we need to be sceptical about this growth, and what it might ultimately mean. Notwithstanding some of the more general and established concerns about the utility of GDP (and GVA) as measures of growth, and the specific difficulties of measuring an industrial sector mainly comprised of (many unregistered and hard to measure) small- and micro-enterprises, it is something of a moot point that the creative economy might currently be growing when the wider economy on which its success depends remains in deep sclerosis. Further, any reported creative economy growth has tended to be quite concentrated, both geographically and sectorally, mostly benefitting 'core' cities, and occurring in mainstream IT and software sectors that might not even be regarded as 'creative' or 'cultural' in any conventional sense (see Banks, Chap. 2). Growth has been further propped-up by reduced investment in education and training, low wages, heightened inequalities, and the exploitation of (increasingly) limited natural resources and environmental assets. These trends do not suggest an especially robust, stable or 'sustainable' basis for future expansion—not least because of the very real possibility of an intensification of limits on the carrying capacities of the economy, society and environment.

While many policy-makers have chosen to ignore these uncomfortable truths, it is of course the case that some mainstream ecological approaches (such as 'sustainable development', 'inclusive growth', 'green growth' and so on) have been identified as potentially adaptable to the creative economy (Florida 2017; OECD 2012; RSA 2017). Yet while such approaches might advocate a 'responsible' economic growth, they do so within an established capitalist economic framework, 'despite overwhelming evidence of its unsustainability and the inequalities inherent in it' (Kothari 2018, p. 251). Such approaches fail to countenance that we might have reached a critical juncture where the old model seems increasingly untenable:

> … we cannot *have* growth much longer: ecological constraints in terms of resources scarcity and a sinking global absorption capacity increasingly reduce the margin of profitability on investments. Industrialised countries may have reached a threshold at which the feasible growth rates no longer secure employment, social mobility and welfare (…) The enforced creation of additional economic value at the same time creates disvalues as a counterpart—for example more social inequality, isolation and new dependencies' (Rosa 2018, p. 7).

Thus, while we agree the creative economy has an important role to play in global economic futures, we need to challenge the uncritical acceptance of its more banal and upbeat projections. Not only does creative economy discourse have a tendency to (over) promote the creative economy as an unproblematic solution to systemic economic crisis, it also romanticizes the social affordances of creative, digital and AI technology, glosses over social and workplace inequalities, and pays only lip service to ideas of environmental 'sustainability'. Indeed, it tends to be short-term *economic* sustainability that is prioritized over concerns for, say, equal opportunity, fair pay or environmental care. Such is the nature of economic policy, one might reasonably argue, yet to cling doggedly to the idea that the cultural and creative industries will help accelerate *return* to a pre-crash world of unlimited expansion and growth, is to deny the realities of a more complex and unsettling present, one where a reprise of the kinds of carbon-fuelled growth witnessed in the twentieth century is no longer a realistic option. Even on its own economic terms, the creative economy fails to account for some of the very real economic, social and environmental *limits* now being posed to a model based on compound growth, and the unfettered exploitation and extraction of people and planetary resources.

By the terms of its own rhetoric and publicity, the creative economy imagines itself as both cutting-edge and forward-thinking. It projects itself as a dynamic means to future prosperity, based mainly on rebooting a capitalist model of market- and technology-led expansion inherited from the past. But in the midst of an unfolding set of economic, social and environmental emergencies, how sustainable is such a model? And what alternatives might there be to effect change? The chapters in this book seek to address these, and similar, questions.

1.4 The Chapters

Each of the chapters is self-contained, and offers its own distinct argument or case study designed to address the theme of cultural industries and environmental crisis. There are, inevitably, some shared concerns that feature prominently: environmental extraction and resource consumption, waste and pollution, labour and social justice, expropriation, place and gentrification for example, but many of our authors are also dealing with a number of these issues simultaneously. Our intention was to invite different insights from a range of cross- and inter-disciplinary perspectives, drawn from a range of national contexts. The chapters variously showcase expertise drawn from cultural and media studies, cultural policy, human and urban geography, music, sociology, and work and organization studies, and from case research undertaken in the UK and wider Europe, North and South America, as well as research with a more integrative and global inflection. While each chapter stands alone, it also speaks to a paramount concern—about the possible futures for culture and cultural industries production in a period of accelerating and uncertain environmental (and wider ecological) change.

We begin with **Mark Banks**' chapter which continues the theme of this introduction with a critique of current approaches to cultural production in the creative economy and an analysis of the ways in which policy and governments are failing to address existing and incipient crisis. More importantly, he introduces the idea of 'creative de-growth' as one way of thinking through future economic scenarios for the cultural industries. As in the CUSP project, this looks to a future when cultural resources are decoupled from ideas of capitalist economic growth. As Banks argues, however desirable this might be, the notion of the arts as a potential series of *denials* or *limitations* is something that is fraught with difficulty, given the heritage of the Enlightenment and the lingering impacts of Romanticism that pervade cultural production in the Global North.

Rick Maxwell's chapter, alongside those of Brennan and Caraway, looks at the current and historical practices of some specific forms of cultural industry production. Its starting point is a case study of Eastman Kodak, a US company whose failure, in Maxwell's words, was 'rooted in its contempt for the ecosystems it inhabited'. This frames the subsequent discussion about environmental accounting and reporting, and outlines the risks that companies face if they fail to incorporate green accounting practices. As Maxwell has it, the social liabilities highlighted by green accounting have been largely ignored in the creative economy sector, due in part to the sector's delusion that it is inherently clean and environmentally benign.

Matt Brennan is both an academic and a musician, and his chapter draws on a recent research project entitled 'New Directions in Music and Sustainability Research'. It looks at environmental sustainability as it relates to both live and recorded music, and then focuses particularly on musical instruments, in this case the drum kit, as a case study of a cultural object with an environmental, social and cultural 'footprint.' The chapter alludes also to a crucial wider issue—on what terms might we seek to continue to enjoy pleasurable and meaningful forms of cultural production in an era of profound environmental challenge and change?

Based on her work in Argentina, **Paula Serafini's** chapter picks up some of the ideas of resource 'extractivism' and exploitation raised by Maxwell. The intensive and extensive extraction of natural resources for export, a model that both characterized and plagued many Latin American economies, is considered by Serafini in the case of the cultural industries, where the impact of a creative economy model has been associated with gentrification and the withering of support for local and indigenous cultures. The idea of 'resilience' is flagged as important here and she concludes by looking at proposals for alternative models of cultural production that might provide for a more genuine sustainability in the face of an extraction crisis.

Brett Caraway writes about Amazon, one of the world's largest media companies and largest consumers of energy. This chapter seeks to re-establish the material connections between 'online' companies and the physical world by considering the broad ecological impacts of Amazon's business practices. Looking at shareholder reports and other evidence, Caraway pays particular attention to Amazon's logistics system and its considerable ecological implications. One of his arguments is that any gains in environmental efficiency through e-commerce must surpass the overall increase in throughput associated with the expanded sale of Amazon-supplied goods and services. Yet, for Caraway, Amazon's sustainability initiatives currently not only 'fall far short of reversing the entropic dissipation of resources', they are liable to 'dramatically increase the rates of material throughput' by encouraging and enabling processes of accelerated and unfettered consumption.

Cecilia Dinardi's chapter also looks at Latin America and continues some of the themes raised by Serafini. It examines the extent to which the informal creative economy—which is often located in peripheral urban neighbourhoods and slums—can often provide a genuine alternative to mainstream or purely commercially-led cultural and creative industries. Using examples in Rio de Janeiro and Buenos Aires, she looks at how public support for informal economic projects in arts, culture and crafts may generate both social and cultural (as well as economic) benefits. As she argues, engaging with—rather than denying the existence of—hidden and informal economies of culture and creativity offers a good starting point for public policy to develop more inclusive programmes for the sector, aimed at those traditionally excluded from the economic mainstream. But she cautions that, while policy remains underpinned by a market-oriented logic based on individualistic entrepreneurialism and commercialization, any gains are likely to be precarious.

Continuing the theme of urban cultural production, **Mariangela Lavanga** and **Martina Drosner** draw on their research in the city of Amsterdam. Like Dinardi and Serafini they are interested in the motivations and politics of cultural workers and how these might offer a counterweight to a mainstream market logic. By taking the case of De Ceuvel—a planned workplace for creative and social enterprises—the chapter looks at the aspirations and contradictions of enterprises that claim to put sustainability and the 'circular economy' at the heart of their emerging business models.

Harry Pitts explores the pressures endured by a particular kind of creative worker—those working mainly in the design, branding and advertising industries in the UK and the Netherlands. Here, creative labour is undertaken against a backdrop of constrained spending, economic uncertainty and sectoral competition. Drawing insights from Marxian theory, he shows how the crisis tendencies of capitalism work to undermine the possibility of enduring and meaningful work in the creative sector, but for different workers, and in different and distinctive ways. The chapter concludes by highlighting some of the ways in which freelancers might be seeking to create sustainable ways of working, more resistant to capitalist closure.

Ealasaid Munro's chapter takes as its starting point the idea that cultural policy, as it stands, is—in a Deleuzian sense—'exhausted'. By looking at small-scale cultural and creative production in rural Scotland, she argues that exhaustion is increasingly a dominant characteristic of the experience of working in the cultural and creative industries, and that this can close down the possibility for creativity and expression on both the part of both creatives and audiences. Munro's chapter also explores, however, how personal exhaustion, allied to increasing awareness of *environmental* exhaustion, can also be generative, providing a catalyst or driver to be able to 'do' creative production differently.

The following chapter by **Anthony Killick** and **Kate Oakley** reports directly from the CUSP project. Based on interviews in very different parts of the UK—a hyper-gentrified inner London district, a small rural town and a de-industrialized city in the Midlands—it examines the ideological imaginary of the creative economy and the role it plays in the lives of young aspirant creatives. The chapter argues that place and locality helps distinctively shape understanding of what is possible in terms of creative work and how place consciousness and materiality can intersect with social class in the shaping thoughts of aspiration and limitation around cultural industry careers.

Jonathan Gross and **Nick Wilson** complete our volume with a plea for cultural industry policy makers and professionals to more fully embrace the political opportunities being suggested by emergent theories of a 'Green New Deal' (GND). The chapter suggests that the necessary 'infrastructural revolution' demanded by GND advocates will both require and benefit from a re-invigorated cultural democracy, committed to plural and diverse forms of cultural production and expression. Like us, Gross and Wilson locate the cultural industries at the centre of those processes of sense- and meaning-making that will be so crucial to the ways in which ecological crises will either be exacerbated or challenged.

Throughout this book we have invited all contributors to bring to light some of the ways in which cultural production is implicated in environmental crisis; to show how the seemingly pleasurable and benign worlds of creativity, technology and symbol-making also have their ecological contradictions and costs. Yet at the same time, we also want to help tell a more positive story about what might be done to counteract or alleviate crisis, or to report or imagine other kinds of less exploitative, degraded or diminished ways of making, circulating and engaging with culture. As all our contributors show, while there is certainly much to be concerned about, there are also new possibilities for critique and action, and an already constituted and evolving cultural politics of opposition in the cultural industries themselves. The generative and progressive paradox at the heart of the creative economy is that cultural production not only provides the basis for economic growth and the pursuit of accumulation, but also a frequent source of opposition to these unstable and alienating imperatives. While culture is routinely placed in service to capital, it is also the case that the necessary freedoms afforded cultural workers to ensure spontaneous and innovative creation have a potentially radical and 'transformative value' (Keat 2000, p. 157), since it is often within the practices of cultural industries work and production that people find the necessary resources for building critical capacities to enable them to reflect on the quality and necessity of prevailing conditions. In this respect, it is our hope and aspiration that the cultural industries may come to matter less as engines of economic growth and rather more for their capacity to provide a valued context and resource for evaluating the current order of *life*—including the role played by economic growth in sustaining or supressing such life. Just *how* cultural industries might further contribute to a 'sustainable prosperity'—a good life for all, decoupled from slavish adherence to accumulation and the growth imperative—is the fundamental question we seek to raise and encourage others to explore.

References

Banks, M. 2018. Creative economies of tomorrow: Limits to growth and the uncertain future. *Cultural Trends* 27 (5): 367–380.

Bazalgette, P. 2017. The independent review of the creative industries. London: DDCMS. https://www.gov.uk/government/publications/independent-review-of-the-creative-industries. Accessed 19 Feb 2020.

BMWi. 2017. *Cultural and creative industries monitoring report*. Berlin: Federal Ministry for Economic Affairs and Energy.

Canada Heritage. 2017. Creative Canada policy framework. Government of Canada. https://www.canada.ca/en/canadian-heritage/campaigns/creative-canada/framework.html. Accessed 20 Feb 2020.

Caraway, B. 2017. Literal media ecology: Crisis in the conditions of production. *Television & New Media* 9 (5): 486–503.

DDCMS. 2017. *UK digital strategy*. London: DDCMS.

Demos, T.J. 2018. The agency of fire: Burning aesthetics. e-*Flux Journal #98*. https://www.e-flux.com/journal/98/256882/the-agency-of-fire-burning-aesthetics/. Accessed 19 Feb 2020.

EU. 2018. A new European agenda for culture: Background information. Brussels. https://ec.europa.eu/culture/news/new-european-agenda-culture_en. Accessed 20 Feb 2020.

Florida, R. 2017. *The new urban crisis*. London: Oneworld.

France Créative. 2013. Création sous tension 2e Panorama de l'économie.

Gabrys, J. 2011. *Digital rubbish: A natural history of electronics*. Ann Arbor: University of Michigan Press.

Jackson, T. 2016. Towards a sustainable prosperity. CUSP Blog https://www.cusp.ac.uk/themes/prosperity/towards-a-sustainable-prosperity/. Accessed 23 Feb 2020.

———. 2018. The post-growth challenge: Secular stagnation, inequality and the limits to growth. CUSP Working Paper 12. https://timjackson.org.uk/the-post-growth-challenge/. Accessed 23 Feb 2020.

Jacobs, M., and M. Mazzucato, eds. 2016. *Rethinking capitalism: Economics and policy for sustainable and inclusive growth*. Chichester: Wiley-Blackwell.

Kallis, G. 2018. *Degrowth*. Newcastle: Agenda Publishing.

Keat, R. 2000. *Cultural goods and the limits of the market*. Basingstoke: Palgrave.

Kothari, A. 2018. Towards radical alternatives to development. In *The good life beyond growth: New perspectives*, ed. H. Rosa and C. Henning, 251–262. Routledge: Abingdon.

Maxwell, R., and T. Miller. 2012. *Greening the media*. Oxford: Oxford University Press.

———. 2017. Greening cultural policy. *International Journal of Cultural Policy* 23 (2): 174–185.

Oakley, K., and J. Ward. 2018. The art of the good life: Culture and sustainable prosperity. *Cultural Trends* 27 (1): 4–17.

Oakley, K., M. Ball, and M. Cunningham. 2018. Everyday culture and the good life. CUSP Working Paper Series 9. https://www.cusp.ac.uk/themes/a/wp9/. Accessed 13 Mar 2020.

OECD. 2012. Time to act: Making inclusive growth happen. https://www.oecd.org/inclusive-growth/Policy_Brief_Time_to_Act.pdf. Accessed 20 Feb 2020.

Piketty, T. 2014. *Capital in the twenty-first century*. Cambridge: Harvard University Press.

Rosa, H., and C. Henning. 2018. Good life beyond growth: An introduction. In *The good life beyond growth: New perspectives*, ed. H. Rosa and C. Henning, 1–14. Routledge: Abingdon.

RSA. 2017. Inclusive growth commission: Making our economy work for everyone. https://www.thersa.org/globalassets/pdfs/reports/rsa_inclusive-growth-commission-final-report-march-2017.pdf. Accessed 20 Feb 2020.

UNCTAD. 2018. Creative economy programme. http://unctad.org/en/Pages/DITC/CreativeEconomy/Creative-Economy-Programme.aspx. Accessed 19 Feb 2020.

Chapter 2
Creative Economy, Degrowth and Aesthetic Limitation

Mark Banks

2.1 Introduction

An increasing number of critical social science, arts and humanities scholars are turning their attention to studying the ecological impacts of cultural and creative industries production.[1] For example, within media studies, researchers have identified the general necessity of 'greening the media' (Maxwell and Miller 2012), the finitude of the globe and the contribution of media to the destruction of natural resources within it (Cubitt 2015), the exploitative labour and environmental relations of the communications industries (Qiu 2016), the need to develop more 'sustainable journalism' (Miller 2015), and the social burdens imposed by dealing with toxic media and e-waste (Lepawsky 2018). Researchers have sought to imagine new post-capitalist 'media ecologies' where technology companies are made to bear the full cost of their exploitation and despoilment of persons and planetary resources (Caraway 2017). In a similar vein, Murdock (2018, p. 5) has argued for a new 'moral economy of machines' that might help bring media industries to task for their accelerated 'resource depletion, energy use, pollution and waste.' In cultural policy, Maxwell and Miller (2017) have called for further 'greening'—criticising those arts and cultural agencies which do little to take into account the true ecological costs of their aesthetic and organisational endeavours. In seeking positive solutions, theorists of cultural industries work and organisations have proposed alternative economic models that privilege community subsistence and mutual aid over unfettered eco-

[1] Ecological refers generally to the symbiotic and integral relations between people, animals, processes and environment and, more specifically in the case of this project, the relations between *the social, the economic and the environmental.*

M. Banks (✉)
University of Glasgow, Glasgow, UK
e-mail: mark.banks@glasgow.ac.uk

K. Oakley, M. Banks (eds.), *Cultural Industries and the Environmental Crisis*,
https://doi.org/10.1007/978-3-030-49384-4_2

nomic growth, and new research on cultural co-ops and non-profits, and different kinds of 'sharing' economy is rapidly emerging (for example, see Boyle and Oakley 2018; de Peuter and Cohen 2015; Oakley and Ward 2018; Sandoval 2017). Geographers are challenging the existing models of 'creative cities' by proposing more ecologically-sound and sustainable forms of urban cultural production (Grodach et al. 2017; Kagan and Hahn 2011). In environmental humanities and the arts various approaches have been taken to promote the creation of green, sustainable or ecologically conscious art and cultural production (Brady 2016), with some of the most innovative research being undertaken in the field of post-carbon humanities (Szeman and Boyer 2017) and ecological design (Ehrenfeld 2009; Escobar 2017).

While these interventions are highly valuable, none have yet focused specifically on the 'creative economy' or creative economy *policy per se,* as a powerful and distinctive field of knowledge and activity where most of the aforementioned cultural, media and creative industries are theorised, activated and contained. Certainly, none of these researchers have attempted to construct an overarching account of how the creative economy itself, as a distinctive policy object, might benefit from being re-theorised as a set of *ecological*—and not simply economic—endeavours. The aim of this chapter is to help at least part remedy this omission by trying to situate current discourse and thinking around the 'creative economy' in a more expansive theoretical frame, one that seeks to address this lack of ecological concern and—more specifically—challenge the foundational premise of economic *growth* that animates and energises this absence.

2.2 The Growth Imperative

While the creative economy has emerged over two decades as a kind of 'global orthodoxy' (Schlesinger 2017), it has no single or agreed definition. For UNCTAD (2018) the creative economy is:

> … an evolving concept which builds on the interplay between human creativity and ideas and intellectual property, knowledge and technology. Essentially it is understood as the knowledge-based economic activities upon which the 'creative industries' are based.

For the UK's Nesta (2013, p. 34) the creative economy is 'those economic activities which involve the use of creative talent for commercial purposes.' Mostly the term tends to be used as a convenient shorthand to capture the full range of 'cultural and creative sectors' (European Commission 2018)—the total sum of all designated arts, media, and cultural and creative industries. While the meaning and level of uptake of the creative economy concept has varied between nations, as Schlesinger (2017, p. 86) notes, it has proved highly transactable as a 'self-sustaining, self-referential framework of ideas' where the idea of culture as possessing a 'tradeable economic value' is treated as paramount.

Today, the policies of advanced capitalist economies such as the UK, Australia, Canada, France, Germany, and the wider European Union all strongly promote the creative economy as central to the future of national economic *growth*.[2] Highly optimistic growth scenarios (and some real uplift in output) have been widely attributed to creative economy businesses. As well as offering economic benefits, the creative economy comes laden with narratives of cultural dynamism, modernisation and change, allowing governments to associate themselves with positive recovery, and a renascent futurism that suggests advanced capitalism might (at last) begin to shake off the hangover of the global financial crisis. Yet I want to argue that the creative economy's *growth imperative* is one that we need foundationally to challenge, for at least four good reasons.

Firstly, the imperative is highly questionable, even in its own **economic** terms. Many leading economists (and from across the political spectrum) have argued that advanced capitalism has become systemically mired in 'secular stagnation'; that is, a period of sclerosis where flat or zero real growth rates are now being anticipated for the foreseeable future (for example, see Krugman 2014; Stiglitz 2016; Summers 2014). Aggregate demand is static or falling, and being compounded as real productive investment is further sacrificed to support the short-term goals of financialization (Mazzucato 2018). While the '1%' might continue to cream the profits and rents of financialized capitalism, sub-par growth projections for the most advanced economies point to the fundamental weaknesses of capitalism in its late form. It becomes a moot point therefore that the creative economy is allegedly growing when the wider economy on which its success ultimately depends is likely to be stagnant and *non*-growing in the medium to longer term. It's also important to note that any recognized creative economy growth has tended to be highly uneven, both geographically and sectorally, tending to be concentrated in 'global' cities and reliant on the heavy-lifting provided by those computing, software and IT industries that have long dominated (and skewed) the overall statistical estimates of activity in the creative sector. Much of this activity is only 'creative' (and certainly 'cultural') in the very loosest sense. In the UK, the recent renaming of the responsible ministry from the Department of Culture, Media and Sport to the Department of *Digital*, Culture, Media and Sport indicates the weight of priority now afforded the digital and technology-led sub-sectors. Further, even if compound growth were to crank back into life, jump-started by something like a reinvigorated neoliberalism, a return to Keynesian policies, or even, it must be said, a wholly growth-oriented 'green new deal' (see Pollin 2018), we would still need to note that the creative economy, like the economy as a whole, 'is a sub-system of a larger system, the ecosphere, which is finite, non-expanding, [and] materially closed' (Daly 2018, p. 88), and so any economic development premised on continually expanding GDP will eventually hit some hard ecological and bio-spherical limits. Given these constraints to economic growth it now seems more imperative to explore alternative economic models and priorities other than growth—such as those premised on radical forms of redistribution, subsistence and, potentially, restriction.

[2] See, for example, Bazalgette (2017), Canada Heritage (2017), DCMS (2016), DDCMS (2017, 2018), Dossi (2016), European Commission (2018), France Creative (2013), Government of Australia (2013), BMWi (2017) on Germany, and HM Government (2017, 2018).

Secondly, in **social** terms, while some kind of economic growth might be necessary (especially as a means of alleviating poverty in developing economies—for example, see Sternberg 2017), in the advanced capitalist economies the question of what kind (and level) of growth we might now want to encourage has become a more open question. There is convincing evidence to show measures of well-being and life satisfaction have stalled (or reversed) in developed nations even as economic output has grown (Jackson 2009; Mazzucato 2018; Skidelsky and Skidelsky 2012; Wilkinson and Pickett 2009). It could also be argued that if the pursuit of growth is no longer bringing the general benefits and utilities that are being claimed for it, then this might be *even more so* for the creative economy than for elsewhere. One reason for this is that the creative economy relies heavily on those cultural-symbolic commodities—or 'positional goods' as Fred Hirsch (1976) originally termed them—that often provide the basis for social distinction, status and competition. Economic growth can never fully satisfy the demand for positional goods, since newness, novelty and scarcity are at their very essence—the more popular the good, the less special or satisfyingly distinctive it becomes, and so the more the demand for new goods arises in a world where the desire for new commodities now seems limitless. Hyper-consumption is also socially disaggregating since it not only encourages resource depletion but also reinforces damaging social inequalities and helps reduce life to individualized competitions for social power or class status (Lewis 2013). There are limits to how far we might want to accept such arguments, of course—not least because they tend to disclaim many of the progressive social innovations, resources, rewards and pleasures occasioned and enabled by capitalism and growth—but this doesn't disaffirm the necessity of social critiques, nor their now more urgent importance. One of the arguments for developing a new perspective focused strongly on principles of social justice (rather than unobtainable compound growth) is to help tackle the persistent and deep-rooted structural inequalities that have so perniciously excluded and disadvantaged women, ethnic minorities, working-class people and other socially disadvantaged groups in the creative economy workplace (Banks 2007, 2017; McRobbie 2016; Oakley and O'Brien 2016; Saha 2018). As research has repeatedly shown, contrary to popular belief, the creative economy does not so much challenge social inequalities as structurally reinforce them.

Thirdly, when it comes to **environmental** issues, creative economy policy has largely been silent. Partly, this comes from the politically convenient but unfounded myth that the creative industries are naturally 'cleaner' or more ecologically benign (see Maxwell and Miller 2017). In reality, the cultural and creative industries are highly resource-intensive and often flagrantly and dangerously polluting. This includes many of the world's leading digital technology and communications companies, the global film and television industries, publishing, music and the transport, circulation and logistical systems that sustain them (Caraway 2017; Cubitt 2015; Devine 2019; Maxwell and Miller 2017; Murdock 2018). Furthermore, while relative 'decoupling' might frequently occur, as commodity production becomes more efficient, the idea of absolute decoupling—that technological innovations and efficiency-savings in production will be sufficient to ensure that economic growth can continue without fatally damaging ecological capacity—now looks an unrealistic hope (Bellamy-Foster et al. 2009; Jackson 2009). The overall expansion in demand for new use—and new commodities—will far outstrip any resource and energy efficiency savings in the individual production of any one commodity (the so-called *Jevon's Paradox*). This is without even considering the enormous problems of waste and disposal occasioned by digital communications industries, fashion, music, film and television and so on, and the human and environmental costs of disassembly and dumping, mostly imposed on nations of the Global South (Lepawsky 2018). We might reasonably conclude that in its current incarnation, the creative economy—and the policy framework that supports it—is not only failing to deliver on its own promises for clean, inclusive and sustainable growth (Bazalgette 2017; DDCMS 2018), it may be undermining the prospects of societies making the necessary shift towards more sustainable economies, in general.

Finally, there is the further (and perhaps more familiar) **cultural** objection, that the relentless co-joining of the creative economy to ideologies of economic growth has served to crowd out alternative visions of arts, media and culture as shared public goods, or as articulations of collective cultural interests and political identifications—including ones that might form the basis of a *critique* of growth and economism themselves (Banks 2018; Banks and O'Connor 2017; Oakley et al. 2018; Schlesinger 2017). This is a problem since, often, the productive energies contained and released by the creative economy are not necessarily directed towards the priority of economic growth, but instead used to furnish desires for different forms of life that might reject such a priority entirely. Yet, in the policy discourse it's clear that culture (long since re-badged as 'creativity') is now mostly understood in apolitical terms—as a benign and ambient resource first recognised for its 'tradeable value'. For instance, in the UK, policy and discourse on creative economy have become more tightly harnessed to a bland, technocratic agenda whose prevailing imaginary is that culture is that which is 'creative', 'digital', 'immersive' and good for growth, and that it will provide only benefits for both producers and consumers in evolving sets of market or quasi market exchanges (DDCMS 2017, 2018; HM Government 2017, 2018). But, for the reasons I've outlined, this is a broken model and, ultimately, an unsustainable one.

Thus, for a combination of reasons, I would argue it has become necessary to question the orthodoxy that identifies growth as the primary social objective of creative economy policy. Various approaches to 'sustainable development', 'inclusive growth' and 'urban prosperity' have recently been mooted, yet all of these regard continued economic growth in a fundamentally unchanged system as a desirable and achievable norm (Florida 2017; OECD 2012; RSA 2017; Throsby 2017). But, arguably, such assumptions seem increasingly optimistic.

Thus, for the remainder of the chapter I want to explore—as indicative illustration—just *one* of the many possible alternatives proposed to the economic 'growthmania' (Daly 2018) that continues to fuel the conventional economy, and—by extension—the creative economy within it. The particular approach I outline argues that transitioning to a genuinely sustainable world economy must not only involve finding ways to materially restrict and limit certain kinds of economic, social and environmental activity, but to also shrink them in real and substantial terms—in short, to *degrow* them. In the next section, I introduce this resurgent theory of *degrowth*, and assess its current usefulness in the context of developing critical-ecological approaches to creative economy study.

2.3 Towards Creative 'Degrowth'?

Amongst the range of more ecologically-oriented approaches to economy, a theory of degrowth has re-emerged from its diverse origins in 1970s cultural criticism, ecological economics and political and environmental activism. Foundational texts such as *The Entropy Law and the Economic Process* (Georgescu-Roegen 1971), *The Limits to Growth* (Meadows et al. 1972), *The No-Growth Society* (Olson and Landsberg 1974) and *Steady-State Economics* (Daly 1977) paved the way for a recently revived and re-invigorated set of debates between (amongst others) economists, sociologists, environmental scientists and cultural and political theorists (see Alier 2009; D'Alisa et al. 2015; Kallis 2018). Yet, degrowth is not reducible to a single definition or approach, and is perhaps best understood as an interdisciplinary ensemble of practices and perspectives that displays some common emphases, as one of its leading exponents here argues:

> Degrowth signifies, first and foremost, a critique of growth. It calls for the decolonisation of public debate from the idiom of economism and for abolition of economic growth as a social objective. Beyond that, degrowth also signifies a desired direction, one in which societies will use fewer natural resources and will organize and live differently than today. 'Sharing', 'simplicity', 'conviviality', 'care' and the 'commons' are primary significations of what this society might look like (Kallis et al. 2015, p. 3).

It is a characteristic of degrowth approaches to consider 'sustainable development an oxymoron' (Demaria et al. 2013, p. 196) and to argue instead for *actually* sustainable forms of economy that must inherently be more localized, socialized and ecologically-directed. As well as advocating prioritization of a more general ecological consciousness and restraint, a key economic principle of degrowth is that a sustainable scale of resource use is that which can be deployed or allocated at a level of throughput that does not threaten the viability of the ecosystem from which it is obtained. Resources must be extracted at a scale and intensity that is renewable or regenerative, and should not generate wastes or externalities beyond the absorptive capacity of the local system or system as a whole. Furthermore, degrowth places strong emphasis on social justice issues, arguing that significant benefits might be gained from emphasising greater equality and social equity in the social distribution and allocation of already existing (as well as future) resources and opportunities (Kallis et al. 2015). Of further concern here is intra- and inter-generational justice, meaning that there should be a fairer and more equitable allocation of sustainable resources across existing social populations, and recognition of the need to bequeath socially-just arrangements for future and (where possible) past generations (Cosme et al. 2015). Degrowth perspectives therefore prioritize the kinds of sociabilities that place non-economic aspects of human well-being, mutual care and concern at the heart of economic activity and planning in ways that are not growth-dependent (Daly 2018; Jackson 2009). In short, degrowth proposes a *strong reversal* of the current economic orthodoxy that regards the earth as a limitless or endlessly-renewable resource amenable and resilient to anthropocentric exploitation, with exponents calling for a 'democratically-led redistributive downsizing of production and consumption in industrialised countries as a means to achieve environmental sustainability, social justice and well-being' (Demaria et al. 2013, p. 209).

While such a radical agenda might appear complementary to a more ecologically-oriented approach to creative economy thinking, its notable that degrowth thinking has not yet been substantively applied to the cultural or creative industries or the wider creative economy in which they sit. I'd suggest there are three main reasons for this:

- Firstly, and coming from within degrowth theory itself, there appears some reluctance or uncertainty about theorising art, culture or creative production (see D'Alisa et al. 2015). This might be because of negative associations of such practices with liberal or advanced consumer capitalism, or because such activities might be seen as secondary to the more 'serious' business of conceiving of the kinds of austere living that might be required under conditions of environmental degradation, restriction or scarcity. It might be that art and culture are regarded as non-essential or mere ephemera. Yet, in its cultural quietism, degrowth theory reinforces a somewhat traditional, political and scientific prejudice against recognising the necessity of locating art, media and culture (and cultural studies) at the heart of the social formation. While the many existing approaches to 'green' cultural production or 'ecological art' (for example, Brady 2016) might suggest good potential for linkages to degrowth thinking, there is currently no developed theory of culture or creative production contained with any of the (albeit, still re-nascent) degrowth theories or frameworks. This is to the detriment of degrowth theory, generally, in my view. Even if a degrown world eventually emerges, it will both require and demand different forms of cultural and creative production—and we therefore need a theory of culture, creativity and technology, as well as art and aesthetics, in any envisaged degrowth scenarios.
- Secondly, discussion of 'degrowth' (or even close counterpart concepts such as 'post-growth') has not been prominent in the creative economy partly because the animating energies and imagination of the creative economy are now so entirely *congruent* with a growth-led agenda. The creative sector has become so widely idealised as new, novelty-driven, cutting-edge and technologically innovative, that to suggest the need for some kind of restriction or limitation on its activities is to propose an economic heresy. As Philip Schlesinger (2017, p. 1) has remarked, it's actually become quite difficult '*not* to talk approvingly and largely uncritically' (about the creative economy and its growth priorities, in public policy fora. In the UK, for example, the current Industrial Strategy and Creative Industries Sector Deal remains hitched to a growth agenda that only superficially supports the most conventional of sustainable development

principles, and evidences little enthusiasm for any kind of strong ecological programme or green industrial strategy.
- Thirdly, despite the long existence of oppositional or ecological art, most of the conventional ideas of creativity, aesthetics and cultural production that underpin the creative economy remain premised on an array of ideas rooted in enlightenment, liberal and modernist notions of innovation, progress and growth, strongly enabled by carbon and resource-rich economies, and activated by autonomous individuals who have tended to prioritise the pursuit of personal 'creative freedoms' over other social priorities. The cultural and creative industries are so deeply-rooted in ideologies of unfettered personal expression and free creative agency, they might appear strongly antithetical to any politics that seems premised on shrinkage, limitation or restraint or on socially collectivized and communal—rather than individual—responses to ecological crises of various kind.

These latter two points hint most at what I think is a real problem for importing any kind of 'degrowth' or principles of restriction into creative economy thinking—the difficulty in breaking down the association of degrowth with some potentially unappealing forms of austerity and *limitation* (Latouche 2012; Roth 2017).[3] To any given artist or cultural producer, taken at face value, degrowth might simply appear as a kind of hair-shirted asceticism or exercise in joylessness—a difficult sell to anyone demanding the right to 'freely' create for an audience of diverse publics with an infinite panoply of tastes. But it's not merely about degrowth's apparent potential for curtailing artistic freedom or tendencies for gloomily 'propounding a society of sobriety' (Latouche 2012, p. 77). There is also the serious concern that any imposing of ecological limits (and radical exiting of the conventional economy) is a strategy that might just as easily used to support socially oppressive or authoritarian regimes, as challenge them. And as the current rise of right-wing environmentalism has shown, there is nothing intrinsically radical or progressive in 'green' politics (Aronoff 2019). A key challenge, therefore, for conjoining any kind of degrowth and creative economy perspective is to create a progressive political project that can more equitably share and genuinely sustain a world of scarce resources, while also retaining some sense of cultural production that can satisfy human desires for both individual (and more shared) expressive creativity.

2.4 Conclusion

Is degrowth a useful concept for helping us to re-imagine a progressive future for the creative economy? Perhaps. What is certain is that how we choose to address the difficult challenges posed by accelerated ecological crisis is going to be central in

[3] As Steffen Roth writes: 'people find it hard to define their lives as footprints of their omissions and abstinences as long as these forms of *less* are not good for a *more* of something' (2017, p. 1038).

establishing just how far the cultural and creative industries will continue in more or less sustainable and socially-just ways. While the language of degrowth points us towards an important consideration—that some human activities will simply have to stop—it also runs the risk of flattening the discourse into one that fails to take account of what *precisely* we need to desist from and what we might we might also need to expand. This, ultimately, is about distinguishing between the different quantitative and qualitative dimensions of growth (Roth 2017). The percentage rate of GDP growth offers a proxy measure indicative of an unsustainable expansion and resource use that *must* end, but, in contrast, say, life expectancies, clean air, good education and jobs, social welfare supports, technology in service of common goods, are amongst the many things we might argue need to expand and grow. Quality of life and providing majority opportunities to survive and flourish, *including the opportunity to engage in the work of cultural production*, need to be expanded rather than contracted.[4]

Perhaps the immediate aspect of this challenge, at least for academic and activist critics, is establishing how something like a 'degrowth' perspective (or some substantive equivalent) can be given visibility and traction in the cultural and creative industries, especially when such activities are being more readily subsumed under an economistic discourse and regime. One starting point might be to continue to insist on the principle that cultural and creative production is not just economically important, but also vital to human sociability, sustainment and collective well-being. Culture is partly that which allows us to better examine and evaluate the conditions of life, as well as provide the experiential pleasures that make life worth living—and however unpropitious the conditions for doing so, we need to resist the ever more frequent attempts to deny, discredit (and de-fund) this principle. Beyond this, the broader political aim, in my view, must also be to theorize and enact ecological perspectives that aim to guarantee a future sustainability for cultural production, but that *also* retain some commitment to expressive aesthetics, sensory human pleasures, and social and political values that makes culture so creative, distinctive and meaningful. If we think of this, as I do, as potentially a kind of *creative degrowth* we might begin to imagine a form of ecologically-sustainable production that strives positively to accommodate (rather than deny or frustrate) human needs and desires for art and aesthetics, creativity and culture. Yet, perhaps, given ecological pressures and crises, 'creative degrowth' would not simply be a restricted version of 'business as usual' or 'doing the same with less'—it would involve devising fundamentally new relations to aesthetics, creativity and cultural production more suitable for the conditions of ecological crisis present and to come, a form of post-carbon aesthetics, that doesn't rely on the toxic infrastructures of an existing social apparatus.

[4] In a professional and not just amateur sense, degrowth perspectives tend to place cultural production firmly in the realm of leisure, often content to imagine (in Marx-like fashion) post-work utopias where people can enjoy 'being creative and spending unnecessary and unproductive time in experimenting with beauty [where] the right to philosophize, to play, to sing or to write will be communal' (Kallis 2018, p. 121). This is fine, but cultural production is not yet theorized as part of a vibrant and functioning 'degrown' or post-growth *economy*.

Developing a theory of something like 'creative degrowth' (or perhaps 'creative post-growth' or a radical cultural sustainment) might therefore have three key dimensions:

- The production of new theoretical approaches to art and aesthetics, creativity and cultural production better attuned to a creative economy of more resource-restricted and ecologically at-risk societies;
- The provision of theoretically-informed models for new kinds of creative economy organization, work and employment where activity is not predicated on (unsustainable) economic growth but on post- or degrowth thinking and the fostering of a sustainable human well-being and prosperity;
- The enabling and enacting of ecologically renewable and sustainable cultural production that also positively strives to accommodate *human sensory pleasures, needs and desires* for art, creativity and culture, sufficient to ensure the shared and collective production of creative goods of diverse aesthetics, design and value.

From this starting point, we might start to better imagine and enact how creative art, media and cultural production might actually look, feel or sound like in our ecologically-challenged present and futures to come. Under such terms, the creative economy might be seen as a site for 'growth' of a different kind—as a potential investment in an expansion of economic sustainment, a more collective and creative sociability and flourishing, and as a source of social and cultural values that might assume no subordinate status to economic growth and profit-making.

There is much conceptual (and practical) work that creative economy academics and practitioners will need to contribute here—and no easy answers. Those who write about the cultural and creative industries, and the creative economy more widely, will need to be more explicit about the kinds of economies, social relations and cultural expressions we would choose to support, and, ultimately, the kind of physical and immaterial world(s) we might want people to inhabit (Levitas 2017). We will also need to think again about the 'goods' of creative degrowth, or post-growth, or the unfolding conjuncture, and work to ensure these goods are more fully socialized in a dual sense; that is, made more evenly and equitably redistributed in terms of benefits and reward structures, as well as conceived in more holistic terms that attribute equivalent significance to the diversity of social, environmental and cultural values and imaginaries that sit alongside those economic priorities that have come to so perniciously (and now ominously) dominate our conventional understandings of growth, prosperity and the good life to come.

References

Alier, J. 2009. Socially sustainable economic de-growth. *Development and Change* 40 (6): 1099–1119.

Aronoff, K. 2019. The European Far Right's environmental turn. *Dissent*, May 31. https://www.dissentmagazine.org/online_articles/the-european-far-rights-environmental-turn. Accessed 20 Jan 2020.

Banks, M. 2007. *The politics of cultural work*. Basingstoke: Palgrave Macmillan.
———. 2017. *Creative justice: Cultural industries, work and inequality*. London: Rowman and Littlefield International.
———. 2018. Creative economies of tomorrow: Limits to growth and the uncertain future. *Cultural Trends* 27 (5): 367–380.
Banks, M., and J. O'Connor. 2017. Inside the whale (and how to get out of there): Looking back on two decades of creative industries research. *European Journal of Cultural Studies* 20 (6): 637–654.
Bazalgette, P. 2017. The independent review of the creative industries. *Department for Digital, Culture, Media and Sport*. https://www.gov.uk/government/publications/independent-review-of-the-creative-industries. Accessed 20 Jan 2020.
Bellamy-Foster, J., B. Clark, and R. York. 2009. The Midas effect: A critique of climate change economics. *Development and Change* 40 (6): 1085–1097.
BMWi. 2017. *Cultural and creative industries monitoring report*. Berlin: Federal Ministry for Economic Affairs and Energy.
Boyle, D., and K. Oakley. 2018. *Co-operatives in the creative industries*. Manchester: Co-operatives UK.
Brady, J. 2016. *Elemental: An arts and ecology reader*. Manchester: Gaia Publications.
Canada Heritage. 2017. Creative Canada policy framework. Government of Canada. https://www.canada.ca/en/canadian-heritage/campaigns/creative-canada/framework.html. Accessed 20 Jan 2020.
Caraway, B. 2017. Literal media ecology: Crisis in the conditions of production. *Television & New Media* 19 (5): 486–503. https://doi.org/10.1177/1527476417712459.
Cosme, I., R. Santos, and D. O'Neill. 2015. Assessing the degrowth discourse: A review and analysis of academic degrowth policy proposals. *Journal of Cleaner Production* 149: 321–334.
Cubitt, S. 2015. Integral waste. *Theory, Culture & Society* 32 (4): 133–145.
D'Alisa, G., F. Demaria, and G. Kallis. 2015. *Degrowth: A vocabulary for a new era*. London and New York: Routledge.
Daly, H. 1977. *Steady state economics*. Washington, DC: Island Press.
———. 2018. Ecologies of scale: Interview by Benjamin Kunkel. *New Left Review* 109: 81–104.
DCMS. 2016. *The culture white paper*. London: Department for Culture, Media and Sport.
DDCMS. 2017. *UK digital strategy*. London: Department for Digital, Culture, Media and Sport.
———. 2018. *Culture is digital*. London: Department for Digital, Culture, Media and Sport.
Demaria, F., F. Scheider, F. Sekulova, and J. Martinez-Alier. 2013. What is degrowth? From an activist slogan to a social movement. *Environmental Values* 22: 191–215.
Devine, K. 2019. *Decomposed: The political ecology of music*. Cambridge, MA: MIT Press.
Dossi, S. 2016. The creative Europe programme: European implementation assessment. European Parliamentary Research Service. http://www.europarl.europa.eu/at-your-service/en/stay-informed/research-and-analysis. Accessed 20 Jan 2020.
Ehrenfeld, J. 2009. *Sustainability by design*. New Haven: Yale University Press.
Escobar, A. 2017. *Designs for the pluriverse*. Durham and London: Duke University Press.
European Commission. 2018. A new European agenda for culture: Background information. Brussels. https://ec.europa.eu/culture/news/new-european-agenda-culture_en. Accessed 20 Jan 2020.
Florida, R. 2017. *The new urban crisis*. London: Oneworld.
France Créative (2013) *Création sous tension 2e Panorama de l'économie de la culture et de la creation en France*. http://francecreative.org/.
Georgescu-Rogen, N. 1971. *The entropy law and the economic process*. Cambridge, MA: Harvard University Press.
Government of Australia. 2013. Creative Australia: National cultural policy. http://apo.org.au/node/33126. Accessed 20 Jan 2020.
Grodach, C., J. O'Connor, and C. Gibson. 2017. Manufacturing and cultural production: Towards a progressive policy agenda for the cultural economy. *City, Culture and Society* 10: 17–25.

Hirsch, F. 1976. *The social limits to growth*. London: Routledge.
HM Government. 2017. *Building our industrial strategy*. London: HMRC. https://www.gov.uk/government/uploads/system/uploads/attachment_data/file/664563/industrial-strategy-white-paper-web-ready-version.pdf. Accessed 20 Jan 2020.
———. 2018. *Creative industries sector deal*. London: HMRC. https://www.gov.uk/government/publications/creative-industries-sector-deal. Accessed 20 Jan 2020.
Jackson, T. 2009. *Prosperity without growth: Economics for a finite planet*. London: Earthscan.
Kagan, S., and J. Hahn. 2011. Creative cities and (un)sustainability: From creative class to sustainable creative cities. *Culture and Local Governance/Culture et gouvernance locale* 3: 1–2. https://doi.org/10.18192/clg-cgl.v3i1.182.
Kallis, G. 2018. *Degrowth*. Newcastle: Agenda.
Kallis, G., F. Demaria, and G. D'Alisa. 2015. Introduction: Degrowth. In *Degrowth: A vocabulary for a new era*, ed. G. D'Alisa, F. Demaria, and G. Kallis, 1–18. London and New York: Routledge.
Krugman, P. 2014. Four observations on secular stagnation. In *Secular stagnation: Facts, causes and cures*, ed. C. Teulings and R. Baldwin. London: CEPR Press. https://voxeu.org/content/secular-stagnation-facts-causes-and-cures. Accessed 20 Jan 2020.
Latouche, S. 2012. Can the Left escape economism? *Capitalism, Nature Socialism* 23 (1): 74–78.
Lepawsky, J. 2018. *Reassembling rubbish: Worlding electronic waste*. Cambridge, MA: MIT Press.
Levitas, R. 2017. Where there is no vision, the people perish: A utopian ethic for a transformed future. *CUSP Essay No. 5*. http://www.cusp.ac.uk/wp-content/uploads/05-Ruth-Levitas-Essay-online.pdf. Accessed 20 Jan 2020.
Lewis, J. 2013. *Beyond consumer capitalism: Media and the limits to imagination*. Cambridge: Polity.
Maxwell, R., and T. Miller. 2012. *Greening the media*. Oxford: Oxford University Press.
———. 2017. Greening cultural policy. *International Journal of Cultural Policy* 23 (2): 174–185.
Mazzucato, M. 2018. *The value of everything: Making and taking in the global economy*. London: Allen Lane.
McRobbie, A. 2016. *Be creative: Making a living in the new cultural industries*. Cambridge: Polity.
Meadows, D., D. Meadows, J. Randers, and W. Behrens III. 1972. *Limits to growth: A report for the Club of Rome's project on the predicament of mankind*. New York: Universe.
Miller, T. 2015. Unsustainable journalism. *Digit Journal* 3 (5): 653–663.
Murdock, G. 2018. Media materialities: For a moral economy of machines. *Journal of Communication* 68 (2): 359–368.
Nesta. 2013. *A manifesto for the creative economy*. London: Nesta.
Oakley, K., and D. O'Brien. 2016. Learning to labour unequally: Understanding the relationship between cultural production, cultural consumption and inequality. *Social Identities* 22 (5): 471–486.
Oakley, K., and J. Ward. 2018. The art of the good life: Culture and sustainable prosperity. *Cultural Trends* 27 (1): 4–17.
Oakley, K., M. Ball, and M. Cunningham. 2018. Everyday culture and the good life. *CUSP Working Paper No. 9*. https://www.cusp.ac.uk/themes/a/wp9/. Accessed 20 Jan 2020.
OECD. 2012. Time to act: Making inclusive growth happen. https://www.oecd.org/inclusive-growth/Policy_Brief_Time_to_Act.pdf. Accessed 20 Jan 2020.
Olsen, M., and H. Landsberg. 1974. *The no-growth society*. London: W.W.Norton.
de Peuter, G., and N. Cohen. 2015. Emerging labor politics in creative industries. In *The Routledge companion to the cultural industries*, ed. K. Oakley and J.O'. Connor, 305–318. London: Routledge.
Pollin, R. 2018. Degrowth vs a green new deal. *New Left Review* 112: 5–25.
Qiu, J.L. 2016. *Goodbye i-Slave: A manifesto for digital abolition*. Champaign: University of Illinois Press.

Roth, S. 2017. Marginal economy: Growth strategies for post-growth societies. *Journal of Economic Issues* 51 (4): 1033–1046.

RSA. 2017. Inclusive growth commission: Making our economy work for everyone. https://www.thersa.org/globalassets/pdfs/reports/rsa_inclusive-growth-commission-final-report-march-2017.pdf. Accessed 20 Jan 2020.

Saha, A. 2018. *Race and the cultural industries*. Cambridge: Polity.

Sandoval, M. 2017. From passionate labour to compassionate work: Cultural co-ops, do what you love and social change. *European Journal of Cultural Studies* 21 (2): 113–129. https://doi.org/10.1177/1367549417719011.

Schlesinger, P. 2017. The creative economy: Invention of a global orthodoxy. *Innovation: The European Journal of Social Science Research* 30 (1): 73–90.

Skidelsky, E., and R. Skidelsky. 2012. *How much is enough? Money and the good life*. London: Penguin.

Sternberg, R. 2017. Creativity support policies as a means of development policy for the Global South? A critical appraisal of the UNESCO Creative Economy Report 2013. *Regional Studies* 51 (2): 336–345.

Stiglitz, J. 2016. Inequality and economic growth. In *Rethinking capitalism*, ed. M. Jacobs and M. Mazzucato, 134–155. Chichester: Wiley.

Summers, L. 2014. Reflections on the 'New secular stagnation hypothesis'. In Secular stagnation: Facts, causes and cures, eds C. Teulings and R. Baldwin. London: CEPR Press. https://voxeu.org/content/secular-stagnation-facts-causes-and-cures. Accessed 20 Jan 2020.

Szeman, I., and D. Boyer, eds. 2017. *Energy humanities*. Baltimore: Johns Hopkins Press.

Throsby, D. 2017. Culturally sustainable development: Theoretical concept or practical policy instrument? *International Journal of Cultural Policy* 23 (2): 133–147.

UNCTAD. 2018. Creative economy programme. http://unctad.org/en/Pages/DITC/CreativeEconomy/Creative-Economy-Programme.aspx. Accessed 20 Jan 2020.

Wilkinson, R., and K. Pickett. 2009. *The spirit level: Why equality is better for everyone*. London: Penguin.

Chapter 3
Green Accounting for a Creative Economy

Richard Maxwell

3.1 Introduction

Creative products command a growing share of national wealth creation, generating worldwide revenue that reached $2.25 trillion in 2015 (EY 2015). The US Bureau of Economic Analysis and the National Endowment for the Arts (NEA) estimate that arts contributed $763.6 billion to the US economy in 2017, 'more than agriculture, transportation, or warehousing' (National Endowment for the Arts 2018). Meanwhile, international trade in creative goods and services totaled $509 billion in 2015, growing at an average annual rate of 7% between 2002 and 2015, despite a drop of 12% between 2014 and 2015 (United Nations Conference on Trade and Development 2018, pp. 9–10). The US creative industries had a trade surplus of $20 billion in 2017, 'leading with movies and TV programs and jewelry' (NEA 2018), up from a negative trade balance of just a few years earlier.

These indicators herald the economic importance of creative and cultural goods, adding additional lustre to the creative sectors in a world already enchanted by symbolic values that we have come to ascribe to these enterprises and their products. But rising sales of design goods, fashion, and audiovisuals are offset by increasing waste generated by quickly shifting tastes and the fast turnover of products fueled by marketing and advertising. The volume of discarded clothing and furniture, the Ikea effect on fast fashion in interior design, and e-waste are symptomatic of the problems that lie hidden in the creative sector. Before the end-of-life tsunami of wasted and discarded articles, the creative sectors are already deeply enmeshed in ecosystem welfare, drawing from the extractive industries supplying raw materials that go into the capital goods that serve as tools for creative production, then powering their operations with a mix of energy sources, many of which rely on fossil fuels that contribute to global warming (Cooper 2018; Maxwell and Miller 2019; Schor

R. Maxwell (✉)
Queens College CUNY, New York City, NY, USA
e-mail: rmaxwell@qc.cun.edu

K. Oakley, M. Banks (eds.), *Cultural Industries and the Environmental Crisis*,
https://doi.org/10.1007/978-3-030-49384-4_3

2010). Without ongoing vigilance of the upstream and downstream impacts on the environment, the materiality of the creative sectors practically disappears from sight.

It is a problem we know more about on the consumption side, where our frenzied pursuit of immaterial meanings in the things we buy increases the consumption of material resources, placing ever greater harms into the planet's ecosystems and atmosphere. Juliet Schor calls this the 'materiality paradox' (2010, p. 27). Green organizational goals must attend to the consumption side, both domestically and in international trade, as demand steadily rises for creative goods and services, because rising demand for creative goods can pressurize the production side of these sectors making output a higher priority than environmental sustainability. On this score, vigilance provided by green accountancy, broadly defined, can help to ensure that participants don't lose their heads for the sake of art and imagination at the cost of environmental and social wellbeing. It's a tricky business to entangle creative minds with nagging reminders that everything that goes into the production of goods with highly symbolic value has a deeply material impact.

This chapter examines the problems and promises of greening the creative economy. The first section is about Kodak (and is a revised and extended version of a polemic that appeared in Maxwell and Miller 2018). It's a cautionary tale from the twentieth century, when governmental restrictions on industrial pollution were long overdue in addressing ecological decline, lax in enforcement, or non-existent. It provides a kind of baseline for thinking about the hazards of growth in the creative economy. While Kodak might be categorically different from the present-day sectors comprising the digitalized creative economy, the story of their rise and influence over one of the most successful cultural industries in the last century provides an instructive tale for enterprises that plan to grow and position themselves as market leaders. Longevity of the creative sectors is also something that urban planners are investing in as a way to tap into the national wealth creation of creative and cultural industries noted above, perhaps hoping to leap into the growing international market in creative goods and services. Any government leader promoting creative economic zones with integrated innovative clusters would do well to know the Kodak story. The last section is an exploration of political-economic and institutional strategies available today for greening creative sectors. While far from exhaustive, this section points toward areas of hope and green prosperity in managing the creative and cultural economy.

3.2 When a Creative Industry Defined and Defiled a Company Town

The story of Eastman Kodak is fairly well known, up to a point. The company set up shop in Rochester, New York in the early twentieth century. It based its business on an innovative process for creating film stock used in still and motion pictures. It practically democratized photography in the US, making cheap cameras and affordable film processing available to a mass market. With the help of Edison's film trust, its motion picture film stock gained an early foothold in the US market, and held off foreign competitors. And by the time the film industry moved to Hollywood, it was the primary supplier of film stock. Toward the end of the twentieth century, it controlled 80% of the market for chemically-based photographic materials (Maxwell and Miller 2012). In the course of the twentieth century, Kodak Park, as its headquarters and physical plant were called, became a major economic force in upstate New York. The company was also a fixture in US culture, with its ubiquitous yellow boxes of film, popular odes to its Kodachrome products, and its centrality to documenting personal and family 'memories'.

They dominated the photography business. They were admired. They cultivated employees and built their leadership from within. They thought they were invincible. And despite business histories that have linked Kodak's failure to the company's underestimation of the threat posed by digital technology (Kotter 2012), they actually spent billions of dollars on new digital camera systems (Gilbert and Bower 2002). It's true, they didn't scale down their capabilities in chemically-based photographic production. It's probably also true that their leadership at the end lacked the foresight and fortitude to restructure and write off losses, to do what IBM or Fujifilm did when faced with the burden of unloading businesses in the midst of the digitalization of everything (Mui 2012). With gross margins of 70% in its chemical photog raphy business, management was in no hurry to drop its legacy units for a digital future (Munir 2012). In the end, company leaders tried to squeeze every last penny out of Kodak's photographic business, until they declared bankruptcy in 2012.

Even though the company basically invented digital photography in 1975, its leaders were uneasy about the potential of 'filmless photography' to undermine its core business in chemically-based film and paper production. Kodak held religiously onto its identity as a chemical company into the 1990s, finally spinning of its Eastman Chemical Company in 1994 (based in Tennessee, this company thrives today). Nevertheless, after a $2 billion investment in research and development, the company found a footing in the new digital camera market and became a dominant player into the 2000s, until smartphones with built-in cameras took over (Shih 2016; Weissmann 2012). The re-tooling problem was a challenge not only because Kodak failed to expand capabilities in microelectronics early enough; importantly, start-ups could establish new digital imaging businesses with off-the-shelf modular components, a new reality that made barriers to entry into the digital camera business almost non-existent. Kodak would become one among equals in the digital world. And compared to Fujifilm, its only real competitor in chemical photographic processing, Kodak couldn't diversify its businesses fast enough to get out of its commanding position in traditional photography. Fujifilm shrank its film business and expanded parts of its chemical operations into pharmaceuticals and cosmetics, among other courses it took during its reorganization (Kmia 2018).

Although these arguments have been central to most accounts of Kodak's demise, they disregard the firm's environmental debts and social liabilities, which had accumulated for decades, measurable in tonnage of pollutants, toxic chemicals, and lives and habitats threatened by poisons seeping into land and waterways around and downstream from Kodak Park. It's hard to argue that academic and industry analysts had no way to know this. When the Environmental Protection Agency (EPA) started publishing data on industrial toxic emissions in the 1980s, Kodak gained unwanted public recognition for its outstanding dirty record. From the mid-1980s to the late 1990s, the same period Kodak was floundering in the new digital photography market, the company had spewed more toxic emissions into the environment than any other corporation in the state of New York (Orr 2014). People in and around the Kodak Park neighborhood in Rochester had suffered atypical levels of exposure to dangerous chemicals, the most dangerous of which was dioxin, a bio-accumulative and carcinogenic chemical, well-known in upstate New York for the Love Canal disaster of the late 1970s (Foderaro 1989).

In 1989, *The New York Times* ran an article describing Kodak Park as 'an industrial tableau of gray smokestacks and thick white plumes' (Foderaro 1989). Residents reported toxic effluents seeping into groundwater and billowing out of smokestacks. Schools were routinely closed. Tests showed ground water and soil contained methylene chloride, acetone, and methanol. Methylene chloride (also known as dichloromethane) is a solvent used in making film stock and is a potential carcinogen (Environmental Protection Agency n.d.a). A Kodak spokesman was quoted in the *Times* article saying 'Residents are questioning our reputation, and we don't like that,' adding that people should have realized that Kodak was a chemical company, not that it 'just made the yellow boxes' (Foderaro 1989).

Kodak's contempt for the ecosystems it inhabited might have informed its decision to continue chemical businesses instead of phasing them out more rapidly. In accounts of its environmental record, Kodak seemed hapless in the face of what it called the 'escalating complexity of compliance with [environmental] regulations and permits' (Lave et al. 1997, p. 3). From 1988 to 1992, Kodak spent between $250 and $350 million annually on environmental expenditures. They paid a $five million fine in 1994 to the EPA and agreed to invest in a major upgrade of its utilities to settle a number of pre-1990 violations. In 1995, Kodak spent $106 million for pollution prevention, environmental and waste management (Ibid., p. 14). If these fines were insufficient motivation for cleaning up their act, the declining support for Kodak in Rochester, the epitome of a company town, should have sounded the alarm. Residents balked when the firm pledged $100 million to look into environmental problems. How could they trust Kodak, after its spokespeople admitted that the company had intentionally released tons of toxic chemicals a year before. By 1998, the company faced a $185 million lawsuit filed by Rochester residents (New York Times 1998). Unsurprisingly, Kodak's compliance record was abysmal (Environmental Protection Agency n.d.b). Their mendacity and anti-environment record had become liabilities that would contribute to the company's ruin and leave New York taxpayers with a huge clean-up bill (Hanley 1990). After all, it was a glorified chemical business operating in a state that makes deals with polluters all the time (see Hudson River and General Electric).

In its annual 10-K report for 1993, Kodak listed some of the ongoing liabilities it faced from environmental expenditures as part of its operational costs (Eastman Kodak 1994). The company made no reference to its decades of toxic practices, choosing instead to play the role of victim of regulatory overreach. They acted as if they were untouchable, despite the public outcry against the environmental violence they heaped on the city and state. Regulators seemed to be treating them kindly, exempting them from deep cuts into their profits that could have forced them to meet clean air and water standards. Further reasons for their complacency became apparent at the end of the Cold War when declassified documents showed their deep connections to the highest levels of government. In a Kodak facility codenamed 'Bridgehead,' the company developed top-secret aeronautics and counter-intelligence projects throughout the 1960s and 1970s, designing lenses and film stock for the CIA and the US Department of Defense for spy satellite technology (Quinn 2017). From 1966 until its removal in 2007, Kodak had a small reactor with a cache of weapons-grade uranium in a basement of one of its buildings used for materials testing (McConnell and Todd 2012). When news of this reactor emerged in 2011, local Rochester officials reported having no clue it had been there, reflecting the risk and regulatory weirdness of the discovery. A senior researcher at the Center for Nonproliferation Studies in Washington, was baffled: 'It's such an odd situation because private companies just don't have this material,' he said. A Kodak spokesperson couldn't find any record of Kodak telling the public, local police, or fire and hazardous materials officials (Orr 2012; Eastman Kodak Company 2008).

Local residents continued to bear the worst of the toxic legacy at Kodak Park. In 2014, the Department of Justice's US Attorney's Office for the Southern District of New York and the US Environmental Protection Agency settled with Eastman Kodak in an agreement to have the bankrupt company pay its environmental debt to Rochester, Kodak Park, the Genesee River, and for its pollution in Superfund Sites in New York and New Jersey (United States Department of Justice 2014). But the Justice Department and the EPA cut a sweet deal for Kodak, ensuring that New York taxpayers would foot half the bill (Sierra Club 2013). Kodak would be responsible for a paltry \$50–100 million with additional payments of \$2 to \$five million for clean-up and other forms of mitigation. By some estimates, it normally costs upwards of \$100 million to clean up a single toxic waste site, and Kodak had multiple sites at its business park and environs. Needless to say, environmentalists and local activists were appalled by the government's settlement agreement with New York's biggest and least trustworthy polluter (Orr 2013).

And clean-up continues to this day, both of Kodak's history and the environment it carelessly polluted for decades (New York State Department of Environmental Conservation n.d.). In a final display of its polluting ways, the last coal cars rolled into Kodak Park in March 2018. That's right, coal cars. Kodak had powered its operations for over a century using fossil fuels. The plant was one of three remaining coal-fired plants in New York State, at one point generating enough electricity to power the equivalent of 200,000 homes. How it was allowed to continue to spew toxic carbon emissions after its bankruptcy and settlement with the Department of Justice boggles the mind, but here again regulators showed their deference to the company. Historically, the Kodak plant was 'one of the most polluting in New York state—allowed to have less robust pollution control equipment than utility plants because it was an industrial, not utility, boiler' (Sharp 2018).

3.3 How to Account for a Green Creative Economy

Don't act like Kodak. And become acquainted with three vectors of institutional activity where green accounting for the creative economy is either in place or is making inroads: via professional organizations, national state tax and grant initiatives, and city planning.

At the professional level, there are non-profit and trade organizations that encourage members to follow guidelines for creative productions within their sector. Many of these programmes look to disseminate best practices, such as the craft sector in the UK (Craft Scotland, Creative Carbon Scotland, Craft Northern Ireland, the Arts Council of Wales, Design and Crafts Council of Ireland, and Crafts Council) or the audiovisual and new media sector in the UK (Creative Carbon Scotland 2017). In the UK, the 'albert Programme,' sponsored by BAFTA, provides training and accreditation in green routines for the screen industry (Albert n.d.). In the US, the Producers Guild of America provides case studies and guidance for sustainable audiovisual production with its Green Production Guide (Producers Guild of America n.d.), which also serves as a clearing house for linking producers to suppliers of sustainable and energy efficient equipment and services for the audiovisual sector. Sustainable Media Production Canada offers similar networking and best practices resources (Sustainable Media Production Canada n.d.).

3.4 National State Initiatives

The 2015 Paris accord provides signatory states with benchmarks to reduce carbon emissions on a national level. National states are encouraged to support efforts of all sectors to reduce emissions and implement sustainable practices. Tools available to national states include tax incentives, fines, and other legal instruments. But national governments with the resources to spend on infrastructure projects to meet environmental goals must be willing to embrace unconventional solutions to problems associated with climate change. Forward-looking planners and designers have met with great resistance in the US government, for example, which has been unwilling to spend money on projects with untested results, even after a series of storms devasted Puerto Rico and cities throughout the mainland. We are a long way from a creative economy when governments reject creative solutions to climate change. Governments in general, says one Dutch expert, are 'about security and not about risk … Innovation is about risk and not about security' (Jacobs 2017).

Still, governments have strategies for encouraging small to medium sized enterprises to invest in energy efficient technologies, buildings, and other capital expenditures by using tax codes and grants to incentivize the greening of business facilities. France and the Netherlands use versions of accelerated depreciation, which enables companies to reduce tax obligations for purchases of advanced green technologies. To incentivize the move to greener operations, regulators allow a larger portion of the asset's value to be depreciated early in the life of the asset. This gives organizations a lower tax burden at the beginning of an asset's life, offering a buffer of higher revenue for making the transition to greener technology. Grants that subsidize the costs of consultancy and technical advice for greening businesses are available in Ireland, France, and Germany. Governments can also use procurement policies that stipulate that suppliers of technologies to government bureaucracies meet sustainability standards, an effort to make technology markets and supply chains greener. The majority of OECD countries have some form of green public procurement initiatives (Organisation for Economic Co-operation and Development 2018, pp. 60–64).

Various forms of accelerated depreciation can be found in tax laws of many national states, though these are not always used to promote the greening of industry. But where this tax incentive has been applied it has proven to be a useful way to move energy efficient technologies into small and medium sized businesses that lack the capital to make the huge investment in green facilities. As one green accountant put it, 'in order to de-carbonize the economy globally, the accelerated depreciation incentive has to work for the average struggling business on every continent. We cannot depend on goodwill, young entrepreneur/founders, or pioneer companies to make this right' (Larkin 2014). Companies already investing in carbon reductions would find accelerated depreciation a further incentive to advance their green strategies in, for example, the build out of energy efficient global data centers, known as hyper-scale data centers (Maxwell and Miller 2019). Even large companies forced to buy carbon credits because they won't invest in green technologies to lower their emissions could also be persuaded to move away from their dirty practices by these kinds of green tax incentives.

Accelerated depreciation is a wartime invention, beginning with the First World War and expanded during World War II, in an effort to hasten the move of private manufacturers to war-related production. In the post-war period, it was used to stimulate reconstruction projects. In more recent versions, accelerated depreciation has been used as a green tax incentive for purchasing energy efficient technologies. This remains one tool that governments can use to speed up the move to a green economy (Koowattanatianchai et al. 2019, pp. 7–18). And the analogy of a wartime emergency funding is not lost on green accounting specialists who see the eco-crisis as a global level threat demanding that governments invest now in the large-scale mobilization of industry to meet the challenge:

> Accelerated depreciation is a solution that the G20 can implement. We have a global climate and environmental emergency that is poised to destabilize the world's economies and governments. Here is a relatively simple, uncomplicated solution. World leaders, unite! Do something constructive together on behalf of all the world's peoples and businesses! (Larkin 2014).

3.5 Cities

When cities seek to develop the creative economy, they need to do more than subsidize rents, abate taxes, and hope the next big digital platform will come. City leaders need to have a comprehensive and long-range vision to make the built environment sustainable. For example, they can create incentives for creative enterprises to be located near commuter routes with low carbon emissions (safe bicycle routes, low emissions transit systems, etc.) or build new routes in established zones of creative industries. They could invest in plans for net-zero architecture for creative economic zones, leading the way for new building and renovation strategies for energy-efficient structures. Through various architectural designs suited specifically to the 'climate zone' in which creative and cultural enterprises are to be located, city planners can help create a built environment that eliminates carbon emissions and reduces the overall reliance on the electricity grid. These include heating and cooling systems with zero emissions; waste systems that generate energy; roof design that maximizes surface area for photovoltaic systems; battery storage to power buildings at night; passive light window designs to illuminate interiors and provide warmth in cold months; grid access to off-site renewable energy sources; LED lighting; and super-insulated walls and windows (Hill 2017). Planners might involve creative sectors in so-called resiliency projects in which cities rebuild after years of environmental degradation or following extreme weather events or in preparation for future effects of global warming (Jacobs 2017). Efforts like these have emerged in the San Francisco Bay Area (Resilient by Design n.d.), in the Meadowlands of New Jersey (Jacobs 2017), and recently in New York City (New York City Office of the Mayor 2019).

In a sense, what city planners can do is prepare the built environment for creative economic zones where enterprises can easily plug into sustainable locales to run their operations. These eco-neighborhoods would be attractive to large creative organizations looking for energy savings alone, which would incentivize partnerships with city planners in need of private investment. And if the creative organization is small, municipalities could offer loans or other forms of help to bring smaller enterprises into the green creative zones. For example, they could ensure that small to medium enterprises have access to green experts/mentors and green accounting tools through subsidies and grants to build up environmental expertise, perhaps subsidizing the creation of occupations within organizations identified as 'green officers' (or some name) who are provided time to learn green practices.

One idea gaining traction is municipal investment in the formation of cooperatives comprising a heterogenous mix of creative enterprises—design, new media, fashion, architecture, etc.—with various levels of capitalization that can be used internally for financing of partner cooperatives, creating greater resiliency for participants (Scholz 2016). A fund of this nature could be set up as a kind of bank in the creative economic zone where all participants act as shareholders/stakeholders to guide investment strategies. If modelled along the lines of Spain's Mondragón cooperatives, the creative zones could also work to reduce or eliminate the precarious nature of employment in the sector (Ibid.; Chakrabortty 2018; Lovato 2019). Already about 12% of the US workforce is employed in worker-owned enterprises that have successfully demonstrated the viability and resiliency of the cooperative model (Walsh et al. 2018).

Cities and national governments can also work together to improve the conditions for creative sectors to succeed in the growing international trade in creative goods and services. Interestingly, there is evidence that diversity of migrant origins has been a key variable in growth of exports in creative and cultural goods. One study found a '10% increase in the diversity of immigrants (or a reduction in their concentration) increases creative goods exports by 4%.' The authors concluded that 'Attracting immigrants from a more diverse set of origin countries maybe a way to boost exports of creative and cultural products, which are important for the economic prosperity of developed countries' (Orefice and Santoni 2018).

3.6 Conclusion

The good news is that there's growing interest in environmental accounting and reporting to help guide companies with cost-benefit models that integrate environmental goals into the bottom line. Ways to make routine practices greener within organizations involved in the creative economy include some well-rehearsed tasks: repair, recycle, reuse materials; install LED lights; use 'smart' power strips; turn on your computers' sleep mode; eat locally-sourced meals; install low-flow plumbing; compost waste; eliminate single-use plastics; sign renewable energy contracts with your utility; telecommute; cycle to work; provide bicycles at work for errand runs; bundle shipment orders; use public transport; set printers to two-sided printing and reduce paper use; calculate carbon footprint; set baseline quantities for water, energy, waste and stick to it; help your neighborhood go green by setting up employee service programs; strive for next level aspirations for organizational sustainability; ensure suppliers are certified B corporations, fair trade, or other green businesses; and have your organization certified as a B Corp or any variant that abides by the 'triple bottom line' where funding (or profit, it that's your thing) shares the last line of your income statement with social and environmental gains—the profit, planet, people dictum (Cultivating Capital n.d.).

Institutional practices like these are essential and perhaps the core accounting practices in a green creative economy. But as I have argued, the solutions should not be isolated at the level of the organization. National, regional, and local governments have an important role to play in creating conditions for creative sectors to thrive in a green economy. At their worst, political powers help to create environmental monsters like Kodak. At their best, they agree to mandates like the Paris accord. But they must do more. The urgency of climate science calls for immediate political action at all scales of governance. Let's put these changes into effect so we can avoid the toxic mistakes of the past and prepare a future for the next generation of creatives who are growing up in the midst of the Extinction Rebellion and school climate strikes.

References

Albert. n.d. We are albert. https://wearealbert.org/. Accessed 17 Jan 2020.

Chakrabortty, A. 2018. In an era of brutal cuts, one ordinary place has the imagination to fight back. *Guardian*, March 12. https://www.theguardian.com/commentisfree/2019/mar/06/brutal-cuts-fight-back-preston-dragons-den. Accessed 23 Jan 2019.

Cooper, K. 2018. Fast fashion: Inside the fight to end the silence on waste. *BBC News*, July 1. https://www.bbc.com/news/world-44968561. Accessed 23 Jan 2019.

Creative Carbon Scotland. 2017. The scoping study on carbon reduction strategies for the craft and digital content sectors of the arts and creative industries in Scotland. https://www.creativecarbonscotland.com/wp-content/uploads/2018/02/Carbon-Reduction-in-the-Creative-Industries-Scoping-Report-1.pdf. Accessed 23 Jan 2019.

Cultivating Capital. n.d. Greening offices: 25 tips to get you started. https://www.cultivatingcapital.com/greening-offices/. Accessed 23 Jan 2019.

Eastman Kodak Company. 1994. Form 10-K. http://www.annualreports.com/HostedData/AnnualReportArchive/e/NASDAQ_KODK_1993.pdf. Accessed 30 Mar 2018.

———. 2008. *CFX Radiological characterization report—Californium Neutron Flux Multiplier (CFX)*. Louisville, KY: NEXTEP Consulting Group, Inc.

Environmental Protection Agency. n.d.-a. Facility report. https://echo.epa.gov/detailed-facility-report?fid=110000492173&redirect=echo. Accessed 30 Mar 2018.

———. n.d.-b. Risk evaluation for methylene chloride. https://www.epa.gov/assessing-and-managing-chemicals-under-tsca/risk-evaluation-methylene-chloride-0. Accessed 30 Mar 2018.

EY. 2015. Cultural times: The first global map of cultural and creative industries. https://unesdoc.unesco.org/ark:/48223/pf0000235710. Accessed 23 Jan 2019.

Foderaro, L. 1989. Pollution by Kodak brings sense of betrayal. *New York Times*, March 8. https://www.nytimes.com/1989/03/08/nyregion/pollution-by-kodak-brings-sense-of-betrayal.html?pagewanted=all. Accessed 9 Aug 2018.

Gilbert, C., and L. Bower. 2002. Disruptive change: When trying harder is part of the problem. *Harvard Business Review* 80 (5): 94–101. https://hbr.org/2002/05/disruptive-change-when-trying-harder-is-part-of-the-problem. Accessed 9 Aug 2018.

Hanley, R. 1990. Eastman Kodak admits violations of anti-pollution laws. *New York Times*, April 6. https://www.nytimes.com/1990/04/06/nyregion/eastman-kodak-admits-violations-of-anti-pollution-laws.html. Accessed 9 Aug 2018.

Hill, D. 2017. What it takes to go net-zero. *Architect Magazine—Journal of the American Institute of Architects*, October 5. https://www.architectmagazine.com/practice/what-it-takes-to-go-net-zero_o. Accessed 23 Jan 2019.

Jacobs, K. 2017. Rebuild by design's enduring legacy. *Architect Magazine—Journal of the American Institute of Architects*, April 17. https://www.architectmagazine.com/design/rebuild-by-designs-enduring-legacy_o. Accessed 9 Aug 2018.

Kmia, O. 2018. Why Kodak died and Fujifilm thrived: A tale of two film companies. *PetaPixel*, October 19. https://petapixel.com/2018/10/19/why-kodak-died-and-fujifilm-thrived-a-tale-of-two-film-companies/. Accessed 9 Aug 2018.

Koowattanatianchai, N., M.B. Charles, and I. Eddie. 2019. Incentivising investment through accelerated depreciation: Wartime use, economic stimulus and encouraging green technologies. *Accounting History* 24 (1): 115–137.

Kotter, J. 2012. Barriers to change: The real reason behind the Kodak downfall. *Forbes*, May 2. https://www.forbes.com/sites/johnkotter/2012/05/02/barriers-to-change-the-real-reason-behind-the-kodak-downfall/#7a4ce80f69ef. Accessed 9 Aug 2018.

Larkin, A. 2014. A simple accounting change can make green infrastructure more attractive. *Guardian*, April 8.

Lave, L.B., N. Conway-Schempf, and A. Horvath. 1997. Eastman Kodak case implementation of TQEM at Kodak Park's utilities division. World Resources Institute. http://pdf.wri.org/bell/case_1-56973-186-1_full_version_english.pdf. Accessed 9 Aug 2018.

Lovato, R. 2019. Could co-ops solve income inequality? *Medium,* February 1. https://medium.com/@CraftsmanshipMagazine/could-co-ops-solve-income-inequality-cfd9cb5cac01. Accessed 23 Jan 2019.

Maxwell, R., and T. Miller. 2012. *Greening the media*. New York: Oxford University Press.

Maxwell, R. and T. Miller. 2018. The environmental ruin of Eastman Kodak. *Psychol Today*, April 12. https://www.psychologytoday.com/us/blog/greening-the-media/201804/the-environmental-ruin-eastman-kodak. Accessed 16 Jan 2020.

Maxwell, R., and T. Miller. 2019. *How green is your smartphone?* London: Polity Press.

McConnell, D., and B. Todd. 2012. Kodak confirms it had weapons-grade uranium in underground lab. *CNN*, May 16. https://www.cnn.com/2012/05/15/us/new-york-kodak-uranium. Accessed 30 Mar 2018.

Mui, C. 2012. How Kodak failed. *Forbes*, January 18. https://www.forbes.com/sites/chunkamui/2012/01/18/how-kodak-failed/#2b5f4ba06f27. Accessed 9 Aug 2018.

Munir, K. 2012. The demise of Kodak: Five reasons. *Wall Street Journal*, February 26. http://ocw.smithw.org/mgt360/wsj-kodak.pdf. Accessed 9 Aug 2018.

National Endowment for the Arts. 2018. The arts contribute more than $760 billion to the US economy. March 6. https://www.arts.gov/news/2018/arts-contribute-more-760-billion-us-economy. Accessed 23 Jan 2019.

New York City Office of the Mayor. 2019. Mayor de Blasio announces resiliency plan to protect lower Manhattan from climate change. https://www.nyc.gov/office-of-the-mayor/news/140-19/mayor-de-blasio-resiliency-plan-protect-lower-manhattan-climate-change/#/0. Accessed 23 Apr 2019.

New York State Department of Environmental Conservation. n.d. Eastman Business Park. https://www.dec.ny.gov/permits/97804.html. Accessed 23 Apr 2019.

New York Times. 1998. Rochester parents fret, and sue, over cancer, *New York Times*, March 2, B4

Orefice, G., and G. Santoni. 2018. Exporting creative and cultural products: Birthplace diversity matters! *World Economy* 41 (7): 1867–1887. http://www.cepii.fr/PDF_PUB/wp/2017/wp2017-17.pdf. Accessed 23 Apr 2019.

Organisation for Economic Co-operation and Development. 2018. *Environmental policy toolkit for SME greening in EU eastern partnership countries. OECD green growth studies*. Paris: OECD Publishing. https://doi.org/10.1787/9789264293199-en.

Orr, S. 2012. Did you know? Kodak Park had a nuclear reactor. *Democrat & Chronicle*, May 11 (republished on 18 August 2016). https://www.democratandchronicle.com/story/local/2016/08/18/did-you-know-kodak-park-had-nuclear-reactor/88944080/. Accessed 9 Aug 2018.

———. 2013. Is the old Kodak dump safe? *Democrat & Chronicle*, May 15. https://www.democratandchronicle.com/story/news/local/watchdog/2013/05/15/kodak-radioactive-waste-dump-site/2163033/. Accessed 9 Aug 2018.

———. 2014. Latest toxic release data: Kodak down but still up. Democrat & Chronicle, February 5. https://www.democratandchronicle.com/story/environment/2014/02/05/toxic-release-inventory/5224481/. Accessed 9 Aug 2018.

Producers Guild of America. n.d. Green production guide. https://www.greenproductionguide.com/. Accessed 17 Jan 2020.

Quinn, K. 2017. Rochester's remote sensing history—Kodak's long-secret intelligence legacy carries on at Harris Corp. *Trajectory Magazine,* November 10. http://trajectorymagazine.com/rochesters-remote-sensing-history/. Accessed 9 Aug 2018.

Resilient by Design. n.d. Resilient by design bay area challenge. http://www.resilientbayarea.org/about. Accessed 17 Jan 2020.

Scholz, T. 2016. *Platform cooperativism: Challenging the corporate sharing economy*. New York: Rosa Luxemburg Stiftung. http://www.rosalux-nyc.org/wp-content/files_mf/scholz_platform-coop_5.9.2016.pdf. Accessed 23 Jan 2019.

Schor, J. 2010. *Plenitude: The new economics of true wealth*. New York: Penguin Press.

Sharp, B. 2018. Last coal cars roll into Kodak area, making final delivery as power plant converts to gas. *Democrat & Chronicle*, March 14. https://www.democratandchronicle.com/story/news/2018/03/14/last-coal-cars-roll-into-kodak-park-making-final-delivery-power-plant-converts-gas/418650002/. Accessed 23 Jan 2019.

Shih, W.C. 2016. The real lessons from Kodak's decline. *MIT Sloan Management Review* 57 (4): 11–13.

Sierra Club. 2013. Don't stick NY taxpayers with Kodak cleanup bill. Atlantic Chapter October 16. https://atlantic2.sierraclub.org/content/dont-stick-ny-taxpayers-kodak-cleanup-bill. Accessed 23 Jan 2019.

Sustainable Media Production Canada. n.d.. https://sustainablemediacan.ca/. Accessed 17 Jan 2020.

United Nations Conference on Trade and Development. 2018. *Creative economy outlook: Trends in international trade in creative industries*. Geneva: UNCTAD.

United States Department of Justice. 2014. Manhattan U.S. attorney and EPA announce agreement with Eastman Kodak Company for clean up of Rochester, New York, Business Park and the Genesee River. March 12. https://www.justice.gov/usao-sdny/pr/manhattan-us-attorney-and-epa-announce-agreement-eastman-kodak-company-clean-rochester. Accessed 23 Apr 2019.

Walsh, P., M. Peck, and I. Zugasti. 2018. Why the U.S. needs more worker-owned companies. *Harvard Business Review*, August 8. https://hbr.org/2018/08/why-the-u-s-needs-more-worker-owned-companies. Accessed 23 Jan 2019.

Weissmann J. 2012. What killed Kodak? *The Atlantic*, January 5. https://www.theatlantic.com/business/archive/2012/01/what-killed-kodak/250925. Accessed 30 Mar 2018.

Chapter 4
The Environmental Sustainability of the Music Industries

Matt Brennan

4.1 Introduction

> An army of peaceful guerrillas … showed itself imminently ready to turn back on the already ravaged cities and their inoperable 'life-styles,' imminently prepared to move onto the mist-covered fields and into the cool, still woods. (Hodenfield 1969)

In his review of the Woodstock Music and Art Fair, which took place between 15 and 18 August 1969 and attracted an audience of more than 400,000 to the site of Max Yasgur's farm, *Rolling Stone* reporter Jan Hodenfield highlighted a paradox of post-industrial musical culture: people frequently participate in music as a means to *escape* the undesirable aspects of mass culture ('ravaged cities and their inoperable life-styles'), but popular music simultaneously *constitutes* a form of mass culture ('an army' of 400,000 in the case of the Woodstock audience). When we take part in the production and consumption of music, we are almost always interacting with broader social structures and material infrastructures—along with any environmental impacts arising from such interaction. Hodenfield's review of Woodstock, for example, goes on to describe some of the most visible environmental consequences of the festival:

> The sanitation facilities (600 portable toilets had been spotted across the farm) were breaking down and overflowing; the water from six wells and parked water tanks were proving to be an inadequate supply for the long lines that were forming, and the above-ground water pipes were being crushed by the humanity … Police reported a shortage of ambulances, and those that were available had difficulty getting back to local hospitals through the metal syrup of the traffic jam … In addition to the mountain shortages of food and water and overtaxed toilet facilities, a new problem was developing: mounting piles of garbage … At the end of the three-day event, festival officials estimated that the cleanup of the hillside, campsites and roads would take at least two weeks. (Ibid.)

M. Brennan (✉)
University of Glasgow, Glasgow, UK
e-mail: matt.brennan@glasgow.ac.uk

K. Oakley, M. Banks (eds.), *Cultural Industries and the Environmental Crisis*,
https://doi.org/10.1007/978-3-030-49384-4_4

Things have got both better and worse since Woodstock took place over fifty years ago. Whereas the Woodstock generation was making up the rules of large-scale amplified live music events as they went along, such mass gatherings are now a routine part of the global music economy. The music industries have become increasingly professionalized and efficient in delivering these events, but unfortunately it does not follow that their environmental impact has lessened.

What do the dynamics of the music industries look like from an environmental perspective? This is not a popular line of inquiry for those who advocate for music in the face of cuts to arts funding and education that routinely threaten access to musical opportunities, and when factors such as gentrification and property development threaten the existence of music venues and spaces for cultural performance. Against this backdrop, research into the environmental effects of the music industries can be at best unwelcome—and at worst met with open hostility. Drawing on previous work from the fields of ecomusicology (Allen and Dawe 2016; Pedelty 2012) and the political ecology of music (Devine 2015, 2019), as well as my own research, I aim to consider the question above.

As I have argued elsewhere, a world where music does not have an environmental impact is a world without music (Brennan 2020). It is not my intention to ruin one of life's great pleasures—the enjoyment of music—by pointing out its environmental cost. However, if we are to have any hope of addressing the challenge of climate change, we urgently need to become more mindful of the cost of the whole range of consumption behaviours that we usually take for granted, including our participation in music. The rest of this chapter therefore focuses on three of the most important components of the music industries—recorded music, live music, and musical instruments—and considers their current state from the perspective of environmental sustainability. It also offers a critique of the assumption that the growth of these industries is an unquestionable good.

4.2 The Live Music Industry

Up until the coronavirus pandemic, live music was the most economically significant part of the music industries. It first outperformed the recording industry in 2008 (in the UK) and was estimated (again prior to the advent of the pandemic) to achieve a total global revenue of over $30 billion by 2022 (Sanchez 2018). In addition to offering a substantial stream of income for many musicians and businesses, the strength of the live music industry has also contributed to the renewal of urban areas, led to increases in tourism, bolstered national economies, and enhanced the musical lives and communities of many people (UK Music 2019; Webster et al. 2018). Yet the cultural value and commercial successes of live music come with many hidden costs and consequences. Indeed, a live concert is only possible via all the infrastructural relations that underpin it, including transport and travel as well as energy consumption in performance spaces and waste management at festivals.

Given its increasing economic importance in recent years, it is perhaps no surprise that music tourism is being touted as a potential growth area. Karen Bradley (the UK government's Secretary of State for Culture, Media, and Sport at the time) welcomed a 2017 industry report on music tourism by noting that musicians 'drew more than 30 million people to live music concerts and festivals in the UK last year … Four out of ten people going to those events were music tourists … Music and the creative industries are central to our post-Brexit future. Live events in the UK draw visitors from across the globe to spend their money here' (Sanchez 2018, p. 8). Meanwhile, the Sound Diplomacy global music consultancy has also argued for a music tourism growth strategy in a co-authored white paper entitled 'Music is the New Gastronomy':

> Music tourism—as a specific sector of the tourism sector—is emerging, but not wholly understood. Unlike gastronomy or cultural tourism, music tourism is less defined, less organised and as a result, less lucrative. We believe this should change … [We need to] demonstrate the value of music to the tourism sector, and how it can be monetised to increase visitor numbers, hotel stays and other indicators. (Orozco et al. 2018, p. 6)

Proponents making the case for growing the live music sector, and music tourism in particular, tend not to dwell on the environmental impact of the industry. But its impact is real enough: the environmental consultancy Julie's Bicycle once estimated that the UK music industries alone emit at least 540,000 tons of carbon dioxide equivalents (also known as greenhouse gas equivalents) into the atmosphere each year, and that the live music sector accounts for the vast majority—roughly 75%—of those emissions (Bottrill et al. 2008, p. 2).

In recent years, myself and other researchers have proposed the concept of a 'live music ecology' to make sense of the live music sector, subsequently noticing that 'ecology' has also become a commonplace term in music policy documents, replacing previous correlative notions such as creative industry 'quarters' and 'clusters' (Behr et al. 2014, 2016; Holden 2015). However, the goal here is not merely to replace one buzzword with another. An ecological approach to live music should place an emphasis on different aspects of its production: (1) the *materiality* of the buildings in which live music happens and its surrounding infrastructures; (2) the *interdependence* between the actors and infrastructural materials who operate by intention within a music scene as well as those that operate outside of given music scenes (for example, regulators and licensing boards) but which nevertheless have a significant impact on live music; and (3) the *sustainability* of live music culture, where all the factors above contribute to the character and meet the needs of those living in a given musical ecosystem—ideally without, as the World Commission on Environment and Development (1987, p. 16) once put it, 'compromising the ability of future generations to meet their own needs.[1]

[1] Note that my approach to live music ecology stands in contrast to that of Huib Schippers and the Sustainable Futures for Music Cultures Project funded by the Australian Research Council. Schippers does not place a similar emphasis on materiality, interdependence, and sustainability in his theorization of musical ecosystems. He also uses the term 'sustainability' in a different sense, focusing on the sustainability of musical cultures as intangible cultural heritage rather than any

In the case of a concert, for example, audiences must choose how far they will travel to see their favourite artist in concert and what mode of transport they will use to get to the event—decisions that ultimately account for the majority of that event's carbon footprint. If the audience is camping at a festival or staying overnight in a city, their consumption choices, particularly accommodation and subsistence, create additional environmental impacts. The venue for a live music event is also an important factor, particularly whether it is a permanent or temporary space. In the former case, the venue will likely have access to electricity from a power grid. But in the latter case, it may have to install onsite infrastructures such as electricity as well as water and sewage. The venue's proximity to public transport and local residents, as well as its agreements with local authorities and councils on its environmental responsibilities and protocol, will also be important factors. The promoter will liaise with the venue (or in the case of a temporary site, facilitate the installation and demolition of the venue) and work with production suppliers to provide staging, lighting and sound equipment, and so on. In the case of greenfield sites for music, infrastructures such as tents, toilets, trackway, food, waste management, and generators will also have to be procured, each with a range of possible environmental effects that depend on the promoter's choices in sourcing suppliers. Meanwhile, the artist, manager, and agent will agree touring itineraries, where most often planning a route to mitigate environmental impact will lose out to other priorities such as securing the best fee for each performance. This situation is not helped by the fact that artists are generally able to command higher fees through 'exclusivity agreements' in their contract, which state they will not perform within an agreed geographical distance of the concert in question for a specified calendar period. For all of these reasons, the live music industry has clear environmental effects.

It would be unfair to say that the live music sector has completely ignored the issue of environmental sustainability. Indeed, 1960s free pop festivals were often wrapped up in ecological rhetoric (Brennan et al. 2019, p. 255), while in the contemporary era there are numerous agencies and organizations operating as part of the 'green events' sector dedicated to finding ways to reduce the environmental impact of large events of all kinds. In the case of music, the aforementioned British consultancy Julie's Bicycle was a pioneer in this respect and currently works to effect policy changes, encouraging sustainability through engagement between industry and the state as well as conducting 'data collection and research, workshops and training events, quality assurance, capacity building, and thought leadership.[2] Outside the UK, there are a number of comparable organizations across Europe including the Green Music Initiative (Germany), Le Collectif des festivals (France), Green Events (Netherlands), Greener Events (Norway), and Energy Efficient Music Culture (with nine partners in twenty-one countries). Such firms are a quickly growing subsector within the events industry, as evidenced by the annual ADE Green conference, which brings together an average of 500 professionals each year to share best practice on environmental sustainability in the events sector.

concerns relating directly to environmental sustainability (see Schippers 2015).

[2] Julie's Bicycle presentation slides for Fields of Green roundtable workshop, 16 October 2015.

4.3 The Recording Industry

As someone who teaches on a university music industries degree, I often have students asking me for advice on the most effective way to release recorded music. The truth is that there is no steadfast rule anymore, and any answer will depend on the genre in which an artist situates themselves and the type (and size) of audience they hope to reach. Releasing recorded music also raises both economic and environmental questions: if I press my album onto vinyl, would that be a disastrous financial decision? How concerned should I be about arguably unnecessary and inarguably hazardous PVC plastic waste that remains the core ingredient of vinyl records? Would cassette tapes (also plastic) be a good idea, now that they're making a comeback in indie circles? How many people still own tape decks? Is it still worth making (plastic) CDs? What would be the economic and environmental consequences of a digital-only release? Such questions demand an investigation into how the relationship between the economic and environmental costs of recorded music have changed across formats and over history.

The questions above provided the impetus for a collaboration between myself and Dr. Kyle Devine on a research project called 'The Cost Of Music,' where I conducted archival research on how both the units produced and price consumers had been willing to pay for recorded music had changed over time, while Devine investigated the environmental impact of different recording formats.[3] The results were startling to us (Fig. 4.1).

Using the United States as a case study, the infographic illustrates how five economic and environmental metrics have changed over type across seven recording formats: (1) the phonograph cylinder; (2) the 78 rpm disc; (3) the vinyl LP album; (4) the cassette tape; (5) the compact disc; (6) the digital album download; and (7) streaming music remotely from server farms (also known as 'the cloud'). Devine (2015) has previously mapped the history of recorded music into three material eras: shellac, plastic, and data.

The *shellac era* is something of a misnomer: the first viable sound recording and playback device was not the shellac disc, but the phonograph, invented by Thomas Edison in 1877. The cylinder format's peak year of production in the USA was 1907, when 28 million cylinders were manufactured. In today's money (adjusted for inflation), the illustrative price of a cylinder in that year was roughly $13.88. In the nineteenth century, cylinders were made of wax-like substances made from animal fats like beef tallow and processed paraffins like coal-based 'stench-wax'. As the phonograph industry evolved, cylinders were made from all kinds of things—from household tinfoil to whale spermaceti, and finally celluloid and a synthetic called condensate. The gramophone was invented by German-American immigrant Emile Berliner in 1887, and improved on the design of the phonograph by using flat round discs as a storage format instead of cylinders. Discs came in different sizes, but typi-

[3] Economic costs were researched by Matt Brennan with assistance from Paul Archibald (methodology, sources, datasets published in Brennan and Archibald 2019). Environmental costs were researched by Kyle Devine (methodology, sources, and datasets published in Devine 2019).

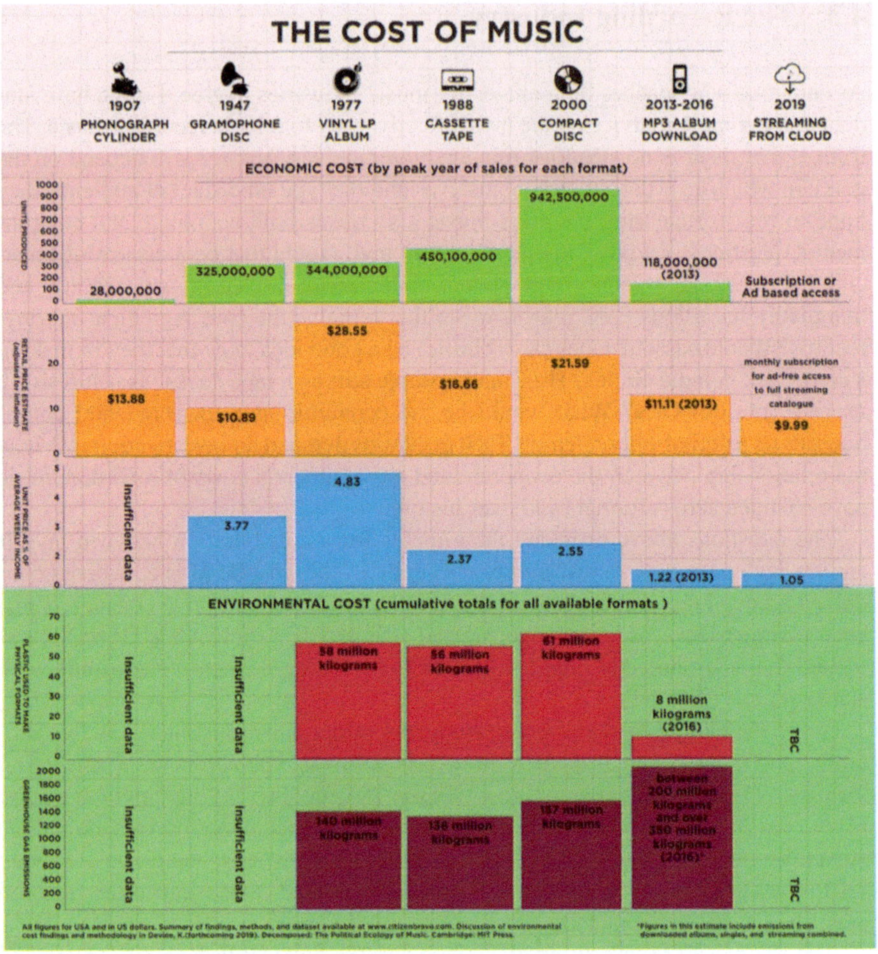

Fig. 4.1 Diagram illustrating aspects of the economic and environmental costs of recording formats over time (for more details see Brennan and Devine 2020)

cally stored roughly three minutes of sound on each side. The 78 rpm disc format's peak year of production in the USA was 1947, when 325 million discs were sold. A key component of gramophone discs is shellac, a natural material produced by lac beetles in India. Once lac was harvested, processed washed, dried, and melted, it was shipped to a disc factory, where various ingredients for 78 rpm discs would be combined with shellac as an important but minority component, while the majority ingredients for discs were locally sourced limestone and slate used as filler material (Brennan and Devine 2020).

The vinyl record moved recorded music away from the acoustic era (where most phonographs and gramophones operated without electricity) into the electrical analog era, but also the *plastic era*. To play music in this era consumers were not only now consuming electricity (an energy cost) but polyvinyl chloride (PVC) plastic. Columbia Records unveiled the 12-in. 33 1/3 LP in 1948, and in 1949 RCA-Victor introduced a rival format, the 7-in. 45 rpm single. The vinyl LP format's peak year of production in the USA was 1977, when 344 million albums were sold. In today's money (adjusted for inflation), the illustrative price of an album in that year was roughly $28.55, equivalent to 4.83% of a US citizen's average weekly salary at the time. The new plastic cost, along with the boom in recorded music manufacturing and sales, had obvious environmental implications. Roughly 58 million kilograms of plastic were used by the US record industry in 1977, the peak production year of the vinyl LP.

The plastic era of recorded music continued with the cassette tape and compact disc. The cassette tape format was developed in Belgium and introduced by the Philips Corporation in 1963. The cassette overtook other tape-based competitors (like the 8-track cartridge and reel-to-reel) thanks to its portability and sound quality. The cassette format's peak year of production in the USA was 1988, when over 450 million cassettes were sold. In today's money (adjusted for inflation), the illustrative price of an album on cassette in that year was roughly $16.66 US dollars. Meanwhile, the compact disc (CD) was a digital-optical data storage format co-developed by Philips and Sony and released in 1982. The compact disc format's peak year of production in the USA was 2000, when over 942.5 million albums on CD were sold. In today's money (adjusted for inflation), the illustrative price of an album on CD in that year was roughly $21.59 US dollars, with roughly 61 million kilograms of plastic used by the US record industry that year.

With the development of the MP3 coding format in 1993, Devine suggests that we enter the *data era* of recorded music. The small file size of MP3 files led to a boom in the distribution of music over the Internet in the mid to late 1990s, when bandwidth and storage were still very restricted. This created a market for MP3 players, and the most successful of these was the iPod, which was first released by Apple in 2001. The MP3 format's peak year of sales in the USA was 2012 for singles, when over 1.39 billion digital singles were legally sold; and 2013 for albums, when 118 million digital albums were legally sold. In today's money the illustrative price of a digital album in 2013 was roughly $11.11 US dollars—or roughly 1.22% of a US citizen's average weekly salary in that year. When downloading takes over, the amount of plastics used by the US drops dramatically to just eight million kilograms by 2016.

The impact of the digital download market is difficult to separate from the impact of streaming, but it is fair to say that, with the advent of streaming, the business model of consuming recorded music gradually changed from being a *commodity industry* (buying copies to own) to a *service industry* (buying access to a temporary experience of music stored in the cloud). For just $9.99, or just over 1% of the current average weekly salary in the USA, consumers now have unlimited ad-free access to almost all recorded music ever released via platforms like Spotify, Apple Music, YouTube, Pandora, and Amazon. Streaming music took off in the mid-2000s, and the key company to make streaming commercially viable was Spotify, a Swedish company that launched in 2008. Streaming platforms use different digital encoding formats; Spotify, for instance, uses a compression format called Ogg Vorbis rather than MP3. We might initially assume that the rises of downloading and streaming are making music more environmentally friendly. But a very different picture emerges when we think about the energy used to power online music listening.

Storing and processing music online uses a tremendous amount of resources and energy—which have a high impact on the environment. (There is an equivalent point to make about waste resulting from playback devices with built in obscelence, of course, but our calculations were focused on recording format units as opposed to playback devices.) It is possible to demonstrate this by translating the production of plastics and the generation of electricity (for storing and transmitting digital audio files) into greenhouse gas equivalents (GHGs). Devine's research shows GHGs of 157 million kilograms in 2000, but by 2016 the generation of GHGs by storing and transmitting digital files for those listening to music online is estimated to be between 200 million kilograms and over 350 million kilograms in the USA alone. The transition towards streaming recorded music from internet-connected devices has resulted in significantly higher carbon emissions than at any previous point in history.

Obviously this is not the last word on the matter. To truly compare past and present, if it were even possible, you would have to factor in the emissions involved in making the devices on which we have listened to music in different eras. You would need to look at the fuel burned in distributing LPs or CDs to music stores, plus the costs of distributing music players then and now. There are the emissions from the recording studios and the emissions involved in making the musical instruments used in the recording process. You might even want to compare the emissions in live performances in the past and the present—it starts to look like an almost endless enquiry. Yet even if the comparison between different eras ultimately came out looking different, our overriding point would be the same: the price that consumers are willing to pay for listening to recorded music has never been lower than today, yet the hidden environmental impact of that experience is enormous.

The calculations in our project were made based on data from 2016, and it is important to note that the carbon impact of streaming would diminish significantly if all aspects of streaming were powered by renewable energy. But as one commenter pointed out in response to an article presenting our initial findings, even with corporations like Google and Apple moving to renewable energy supply, sorting out international energy production issues is a huge challenge:

> rich countries cannot simply press a button to turn on clean energy sources and leave fossil fuels behind. We use far too much energy to take even half of it from renewables. Unless we drastically change our economic systems and our energy-guzzling lifestyles, climate change will send us the way of the dinosaurs and there will be no more music to stream. (Brennan and Devine 2019)

4.4 The Musical Instruments Industry

The musical instruments sector is not typically counted as part of the music industries. This is a strange state of affairs for two reasons. First, the musical instruments manufacturing sector was arguably the first component of the modern music industries to consciously self-identify as 'the music business' from at least the 1890s onward.[4] Second, the devaluation of the recording industry in the early 2000s coupled with the growth of the live music industry led to analysts pluralizing 'music industry' to 'music industries' to better reflect the multiple sectors of industry that monetized musical culture. Yet musical instruments have tended to be ignored even in accounts of the pluralized music industries, a situation made all the more puzzling by the fact that as with live music, the value of the musical instruments industry also showed signs of overtaking the recording industry in the late in the first decade of the twenty-first century (Fig. 4.2).

In the graph, one line indicates global revenues from recorded music according to the International Federation of the Phonographic Industry (2016, p. 9). This includes physical sales, digital sales, performance rights, and synchronization rights. We can see a gradual downward trend from 2007 to 2014 with a slight recovery in 2015. The live music industry is not represented by a globally operating trade association that publishes annual revenue figures, but the *green* dot in the top right corner of the graph represents a recent assessment by the accountancy firm Price Waterhouse Cooper of the value of the live music sector in 2015. It's well above both instruments and recorded music. Interestingly, however, the second line indicates revenues from music products according to the National Association of Music Merchants (2016, p. 54). This includes both traditional instruments and new tech-

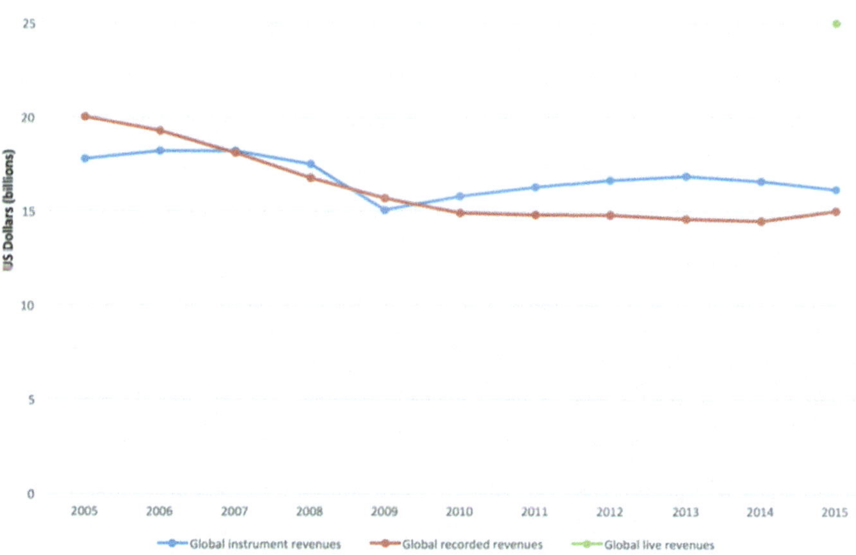

Fig. 4.2 Diagram illustrating comparative global revenues from recorded music (International Federation of the Phonographic Industry 2016), 'music products' (i.e. instruments and music-making technologies, National Association of Music Merchants 2016), and live music (Price Waterhouse Cooper 2015)

[4]This can be evidenced by reading through trade publications such as *Music Trades* from the turn of the twentieth century.

nologies for making music including the following areas: fretted instruments, pro audio and related products, school music, accessories, home pianos, printed music, recording products, percussion, electronic music products, portable keyboards, DJ products, organs, and karaoke devices. (There may be some debate whether all of the music-making commodities listed above should be grouped as 'instruments', but that argument is for another time.) I don't know of a single piece of research that explains the music business in the way the graph suggests we should. Instead, accounts of the music industry have consistently ignored or underestimated the importance of musical instruments and tended to focus on the recorded sector, or more recently, the live sector. Overall it seems clear, however, that it doesn't make sense not to include musical instruments as a key component of the music industry as a whole.

There are several important implications that arise from the inclusion of instruments as a core part of the music industries, but for now I want to focus on what this means when considering the environmental sustainability of music. Inspired by recent work on the ecology of musical instrument making (Allen and Dawe 2016) and Kyle Devine's (2015) aforementioned record on the political ecology of recorded music, I want to propose a framework for making sense of the development of the contemporary musical instrument industry by dividing it into three overlapping historical eras grouped by the principal materials used to manufacture instruments: (1) *renewable and reusable materials* (*wood, metals*); (2) *non-renewable materials* (*plastics and e-waste*); and (3) *data*. I will use the drum kit as a case study to illustrate how this division works in practice, since tracing the history of drum kit manufacturing can reveal much about shifts in material resources used across the musical instrument industry as a whole.

As with other instruments, the drum kit's history is part of a larger narrative of the growth of the musical instrument manufacturing industry over the last two centuries from a small guild-based industry to a globalized mass production industry. From the invention of the drum kit as a modern instrument until the end of the Second World War, its production drew for the most part on *renewable and reusable materials*. In the nineteenth century, drums were mostly made from solid planks of timber wood—maple, mahogany, and walnut were all popular. These planks were then steam-bent into cylinders, after which reinforcing rings or glue rings were added so the shells kept their shape (Nicholls 2008, pp. 16–17). Drum heads, usually made of calf skin, were tightened onto the shell using wooden hoops tensioned by ropes with leather fittings. Cymbals, meanwhile, were (and still are) made of a bronze alloy composed of 80% copper, 20% tin, and trace elements of silver which acted as a catalyst to bond the copper and tin together. The first drum hardware, such as bass drum pedals and snare drum stands, essentially transformed an otherwise disparate collection of drums and cymbals into new hybrid instrument—the early drum kit. By World War I drum manufacturers had started to produce more durable, standardized hardware out of iron and steel. By the end of the 1930s, drum manufacturers were tensioning drums with metal hoops, rods, and fittings, usually chrome-plated steel. They also begin to use laminated plywood instead of solid wood to construct their shells. The acoustic drum kit is still made of mostly wood and metal, and the odd bit of rubber, in the present day. Drum heads, however, are an important exception to this rule, and this brings me to the second overlapping era I want to discuss—the era of *non-renewable and e-waste materials*.

The first non-renewable components of the drum kit were its metal components: cymbals in the nineteenth century were followed in the twentieth century by metal fittings on and accompanying metal hardware typically made of iron, steel, or aluminium and plated in chrome or nickel. Metal snare drums—made of brass, steel, or aluminium—also become popular (especially from the 1960s onward). But most of these metals were also reasonably easy to reuse or recycle, not least because they were usually fairly easy to separate. A more complicated non-renewable material used in drum kits was plastic. As early as 1924, a type of petrochemical-based celluloid plastic known as pyralin was used as a decorative wrap around the shells of drums, and similar plastic wraps only became more popular from the 1930s onward (Brennan 2019). After the war, cheap durable, plastic compounds began to infiltrate most areas of consumer manufacturing, including a polyester film known as Mylar, which like many new postwar production materials had its origins in defense research and development. The dominant plastic drum head manufacturers, Evans and Remo, both began their business in the second half of the 1950s. According to Nicholls, plastic drum heads won over drummers almost completely in a few short years due to their numerous advantages: 'not only were plastic heads stronger than animal skin, they were also weather-proof, heat-proof, consistent, cheaper, and could be produced in vast quantities' (Nicholls 2008, p. 29).

The biggest change in drum kit production during the 1960s and 1970s was without question the rapid rise of Japan as a key centre for drum kit design and manufacturing. Like Germany, Japan was forced to rebuild and rethink across all spheres of industry after the devastation of the Second World War. Non-Western drum manufacturers and, in particular, a trio of Japanese companies—Pearl, Tama, and Yamaha—all began trading in the postwar years. These companies aggressively expanded in the 1970s, not least due to their ability to manufacture drum kits at a cheaper price point thanks to the lower cost of labour in east Asia. The other key consequence, of course, was an increasingly global flow of instruments traveling from continent to continent, thus increasing the carbon footprint of consuming such instruments.

Perhaps the drum kit that best characterizes the non-renewable material era was the electronically synthesized drum kit of the 1980s. A pioneer in this regard was the Simmons SDSV, or Simmons Drum Synthesizer, developed by drummer and record producer Richard James Burgess and collaborators and commercially released in 1981. But electronic drum kits truly began widespread after Roland launched its first V-Drums kit—the TD10—in 1997. As Roland's Product Manager for Drums in Europe, Jules Taberrer-Stewart observed:

> In today's market, we know that electronic drum sales are almost 1:1 for acoustic drums—in the [2015] German market, all acoustic drums and cymbals (excluding percussion) was worth 12.8m Euro, while electronic drum sales accounted for 10.6m Euro in the same year. The industry does not have that information available for all markets in Europe, but the

German market is the largest single market in Europe and other countries will scale similarly (Personal interview, 7 June 2017).

One problem with this shift, of course, is that the majority of drum kits being made and sold today are now made from oil-based plastic and electronics which, like other consumer products, are having an increasingly detrimental effect on the planet.

Related to the rise of electronic drum kits, of course, are drum samples, drum programming and software, which brings us to the *data era* of the instrument. The roots of the data drum kit are in digital sampling. In 1979, Roger Linn created the Linn LM-1 Drum Computer, the first commercially released drum machine to use digital samples of acoustic drums. It was used on hits throughout the early 1980s. By the late 1980s, the replacement or augmentation of the acoustic drum kit by analog drum machines or digital samples of drums was commonplace in pop record production. By the twenty-first century we see the release of dedicated drum replacement software. Unlike streaming recorded music, these software programs do not tend to require the constant remote downloading of data from server farms to operate, but they do of course rely on the wider consumer electronics industry of portable computers, audio interfaces, and other hardware which is both difficult to renew and recycle, and which is characterized by built-in obsolescence and frequent software updates in order to sustain growth.

4.5 Conclusion

Unfortunately, there are no easy solutions to greening the music industries. So what can be done, and by who? Part of the answer to the first question must be a willingness to rethink the rationale for growth of the music industries (the same of course goes for the wider creative and global economy). The second question is equally important. Audiences and listeners clearly have a role to play in their consumption choices. But musicians, both amateur and professional, also shoulder some of the responsibility from their touring practices and travel to their equipment and distribution strategies. Manufacturers, promoters, labels, and technology companies that rely on musical content for their business model will need work with governments to make a transition toward renewable energy supply combined with a reduction in overall energy use.

References

Allen, A., and K. Dawe, eds. 2016. *Current directions in ecomusicology: music, nature, and environment*. New York: Routledge.

Behr, A., M. Brennan, and M. Cloonan. 2014. Cultural value and cultural policy: Some evidence from the world of live music. *International Journal of Cultural Policy* 22 (3): 403–418. https://doi.org/10.1080/10286632.2014.987668.

Behr, A., M. Brennan, M. Cloonan, S. Frith, and E. Webster. 2016. Live concert performance: An ecological approach. *Rock Music Studies* 3 (1): 5–23.

Bottrill, C., G. Lye, M. Boykoff, and D. Liverman. 2008. *First steps: UK music industry greenhouse gas emissions for 2007*. Oxford: Julie's Bicycle.

Brennan, M., and Devine, K. 2019. Music streaming has a far worse carbon footprint than the heyday of records and CDs. *The Conversation* 8 April. https://theconversation.com/music-streaming-has-a-far-worse-carbon-footprint-than-the-heyday-of-records-and-cds-new-findings-114944. Accessed 21 Jan 2020.. Accessed 15 June 2020.

Brennan, M. 2019. *Kick it: A social history of the drum kit*. New York: Oxford University Press.

———. 2020. The infrastructure and environmental consequences of live music. In *Audible infrastructures: Music, sound, media*, ed. K. Devine and A. Boudreault-Fournier. New York: Oxford University Press.

Brennan, M., and P. Archibald. 2019. *The economic cost of recorded music: Findings, datasets, sources, and methods*. Glasgow: University of Glasgow. http://eprints.gla.ac.uk/183249/. Accessed 21 Jan 2020.

Brennan, M., and K. Devine. 2020. The cost of music. *Pop Music* 39 (1): 43–65.

Brennan, M., J. Collinson Scott, A. Connelly, and G. Lawrence. 2019. Do music festival communities address environmental sustainability and how? A Scottish case study. *Pop Music* 38 (2): 252–275.

Devine, K. 2015. Decomposed: A political ecology of music. *Pop Music* 34 (3): 367–389.

———. 2019. *Decomposed: The political ecology of music*. Cambridge, MA: MIT Press.

Hodenfield, J. 1969. Woodstock: 'It Was Like Balling for the First Time'. *Rolling Stone*, 20 September. https://wwwrollingstonecom/music/music-features/woodstock-it-was-like-balling-for-the-first-time-229092/. Accessed 21 Jan 2020.

Holden, J. 2015. *The ecology of culture: A report commissioned by the Arts and Humanities Research Council's Cultural Value Project*. Swindon: Arts and Humanities Research Council.

International Federation of the Phonographic Industry. 2016. *Global Music Report*. https://www.ifpi.org/downloads/GMR2016.pdf.

Music, U.K. 2019. *Music by numbers*. London: UK Music.

National Association of Music Merchants. 2016. *NAMM Global Report*. Carlsbad: NAMM.

Nicholls, G. 2008. *The drum book: A history of the rock drum kit*. Milwaukee: Hal Leonard Corporation.

Orozco, J.G., J. Jones, and S. Shapiro. 2018. *White paper on music and tourism: Music is the new gastronomy*. Bogota: Sound Diplomacy and ProColumbia.

Pedelty, M. 2012. *Ecomusicology: Rock, folk, and the environment*. Philadelphia: Temple University Press.

Price Waterhouse Cooper. 2015. *Outlook insights: An analysis of the Global entertainment and media outlook 2015–2019*. London: Price Waterhouse Cooper.

Sanchez, D. 2018. The live music industry will be worth $31 billion worldwide by 2022. *Digital Music News*, 26 October. https://www.digitalmusicnews.com/2018/10/26/latest-live-music-revenue-31-billion-2022/. Accessed 21 Jan 2020.

Schippers, H. 2015. Applied ethnomusicology and intangible cultural heritage: Understanding 'ecosystems of music' as a tool for sustainability. In *Oxford handbook of applied ethnomusicology*, ed. S. Pettan and J.T. Titon, 134–156. Oxford: Oxford University Press.

Webster, E., M. Brennan, A. Behr, M. Cloonan, and J. Ansell. 2018. *Valuing live music: The UK live music census 2017 report*. Edinburgh: Live Music Exchange.

World Commission on Environment and Development. 1987. *Our common future*. New York: United Nations. https://sustainabledevelopment.un.org/content/documents/5987our common future.pdf. Accessed 21 Jan 2020.

Chapter 5
Cultural Production Beyond Extraction? A First Approach to Extractivism and the Cultural and Creative Industries in Argentina

Paula Serafini

5.1 Introduction

In Argentina, the recent growth and heightened visibility of the cultural and creative industries sits within an economic model that is based on the intensive and extensive extraction of natural resources for export, a model that in Latin America has been termed 'extractivism' (Gudynas 2009; Svampa and Viale 2014). Scholars and activists alike have called for an end of the extractivist model, and an end to a system based on the premise of continuous growth and of capitalist, Eurocentric forms of 'development' (Escobar 2010), which has led to the demise of ecosystems and has curtailed the human and territorial rights of peoples across the region. But while these critiques have recognised the importance of the preservation of local cultures, and involve a broad understanding of art as a space and vehicle for social transformation, the roles of cultural policy and cultural production are often neglected in both critiques of extractivism and proposals for alternative models. At the same time, accounts of creative economies in the region—be that celebratory or critical—are yet to engage with the ecological impact of cultural production, and with the position and functioning of the cultural and creative industries under extractivist models.

This chapter is based on the premise that extractivism is not only an economic model but also a social, political and cultural phenomenon that affects all spheres of private and public life. In other words, we might speak of a logic of extraction. With this in mind, the chapter acts as a first approach to the question: how do the cultural and creative industries mobilize and reproduce a logic of extraction? Taking Argentina as a focus and context of enquiry, I explore this question by adopting the perspective of extractivism to discuss development and the cultural economy. I begin by discussing the concepts of development, extractivism and cultural

P. Serafini (✉)
University of Leicester, Leicester, UK
e-mail: paula.serafini@le.ac.uk

K. Oakley, M. Banks (eds.), *Cultural Industries and the Environmental Crisis*,
https://doi.org/10.1007/978-3-030-49384-4_5

development. I then move on to describe the setup of the cultural and creative industries in Argentina, and the current approach pursued by the government of the City of Buenos Aires. Following this, I analyze a case of state-sponsored and culture-led gentrification in order to exemplify how the logic of extraction is mobilized in the cultural sphere. And, finally, I conclude by suggesting some ways forward in thinking about cultural production beyond extraction.

I propose that this kind of approach is an important intervention because it is only now that scholars of the cultural economy are beginning to look at how 'histories of colonial capitalist development and nation building shape understandings of creativity' (Saifer 2018, p. 25). In the study of Latin American cultural economies, it is paramount to consider how extractive colonialism has shaped cultural imaginaries and practices, as well as economic models. Considering the relationship between cultural production and extractivism can help us develop contextualized critiques to still-emerging cultural economies, while also allowing us to consider how cultural production can play a role in the transition to more sustainable ways of living.

5.2 Development and Extractivism

According to Arturo Escobar (1999), development is a socially produced discourse that contains a geopolitical imagination and is constructed as a historically singular invention rooted in Western, modern ideas of progress (Rey 2009, p. 32). The so-called crisis of development is now decades old. However, the expansion of the extractive frontier under an extractivist economic model in the last two decades has brought about new challenges to regions such as Latin America.

Extractivism is the term employed to describe the model in countries where the economy is heavily reliant on intense extraction of 'natural resources', mainly for export. In Argentina, this takes the shape of an increasingly homogenized model of agriculture for export, the development of mega-projects for non-conventional fossil fuel extraction, and the expansion of open-pit mining. This orientation aims at extracting as much value as possible from the territory, with the claimed objective of funding development and progress. However, according to the United Nations Economic Commission for Latin America and the Caribbean (CEPAL 2014), there is no evidence that extraction-based economies in the region have in fact directed revenues from extractive projects towards significant investments in key areas such as the accumulation of technological capacities, the diversification of production, or the updating of physical and social infrastructure that sustainable economies require. Instead, there is a focus on expanding consumption—of mostly imported goods—which fails to make up for the failings of weakened state institutions (Martín García 2016, p. 87). Furthermore, extractivism leads to the demise of other regional economies such as small-scale farming, the trampling of indigenous land rights, and a series of irreversible ecological consequences, from water pollution to the destruction of ecosystems. Extractivism 'sees territories as commodities, rendering land as for the taking, while also devalorizing the hidden worlds that form the nexus of human and nonhuman multiplicity' (Gómez-Barris 2017, p. 5).

While extractivism is often associated with the extraction of 'natural resources' from the land in mostly rural areas, activists, organised communities and academics in Argentina have begun to draw connections between rural and urban processes under extractivist economies. They argue that there are certain dynamics that are comparable in both contexts, such as the accumulation of capital, the displacement of people, ecological damage, appropriation of public space, and social and institutional degradation. They also point to the effects of increased financialization: while monocrops such as soybean become commodities, so does housing (Viale 2017). These processes are not only occurring in parallel in rural and urban areas, but they are also inter-related. The expansion of industrial agriculture, for instance, has led to mass migration from rural areas towards large urban centres, where the rate of building keeps increasing yet there is no affordable housing, thus leading to an increasing dispossessed population (Ibid., p. 16). The concept of urban extractivism provides a framework for understanding the problems and inequalities of cities not as isolated issues, but as the result of specific development models based on extraction (Vásquez Duplat 2017, p. 9).

When challenging development models, we are faced with the question of what kind of society we want (García Canclini 1987, p. 26). As we stand now in the midst of a global ecological crisis, the international consensus is that development should be sustainable, meaning that it should be based on 'efficient use and equitable distribution of natural resources for long-term, intergenerational socioeconomic well-being' (Maxwell and Miller 2017, p. 175). However, the version of sustainability advocated by multilateral organisations and by several governments still prioritizes economic development over ecological protection, and is still based on a system of continued growth under an unequal capitalist model. A re-thinking of development must entail a move away from extractivism and unlimited growth and towards truly sustainable and democratic economies and societies. With this in mind, I now turn to a critical discussion of the role of culture in development processes

5.3 Development and Cultural Production

Cultural policy became a prominent concern in Argentina and other large economies in Latin America in the second half of the twentieth century, when both Keynesian and Marxist productivist models that were framing planning for development came into crisis. It became apparent that merely economic or political solutions were not enough to deal with social contradictions, population increases and environmental issues. This led both scholars and policymakers to start enquiring about the cultural bases of power and of production, and organizations like UNESCO setting up standards and parameters for cultural development (García Canclini 1987; Szpilbarg and Saferstein 2014). More recently, in his proposal for sustainable development, Hawkes (2001) argued for culture as the fourth pillar of sustainability, highlighting the role of culture in fostering development and progress (Oliva Abarca 2016, p. 4).

But cultural development frameworks have long been, and mostly still are, linked to a Eurocentric project of modernity, and do not fully contemplate the social, cultural and institutional particularities of each country (Rey 2009, p. 24). To this day, for instance, 'creative economy discourse sanitizes and depoliticizes histories of uneven development, and the power relations and structures that continue to shape it' (Saifer 2018, p. 28). In our study of both the global cultural economy and of local markets, especially in the Global South, we must consider 'how ongoing processes of colonization and capitalist accumulation structure the creative economy [and] how dominant knowledges and ways of being shape how the creative economy is understood in policy and practice' (Saifer 2018, p. 24).

In societies where the idea of a cultural economy has only recently been introduced, debates are still being held on whether recognizing and advocating for the role of culture in development and its economic potential is an appropriate path, or whether it poses more dangers than benefits, as creative production can no longer be thought of outside of capitalist economic relations (Szpilbarg and Saferstein 2014, p. 109). A common concern is the dismissal and defunding of practices that are not aimed at generating economic profit (Oliva Abarca 2016; Rey 2009). Also, there is the danger of the massification of culture exacerbating passive consumption as the main form of cultural engagement, and of culture becoming instrumentalized or subsumed to other objectives (Yúdice 2002). And there is concern that promoting the figure of the individual cultural entrepreneur, who creates value through copyright, will lead to a focus on the individual capacity of generating economic value rather than on culture's social value (Szpilbarg and Saferstein 2014, p. 110).

Another concern regarding culture-led development is its potential role in the marginalization and displacement of people. Florida (2002), the champion of the creative class's role in regenerating urban spaces, has recently retracted his enthusiasm, now joining others in claiming that investing in the creative class as a means of solving all social problems has failed (Banks 2018; Florida 2017). Florida acknowledges that these kinds of policies have in fact led to forms of marginalization, displacement and dispossession. Focusing mostly on UK and US cities, he now argues for inclusive, environmentally-friendly policies—such as green transport and investment in health and public infrastructures—that can solve the crises felt by working class and struggling middle class people. However, as Banks points out, his proposals for sustainability still fit within a capitalist programme of growth that continues to ignore the radical changes needed in order to tackle both inequality and the environmental crisis (2018, p. 373).

This leads us to consider the potential environmental effects of emerging and expanding cultural economies, which are not really considered in sustainable development programmes and discourse. There is a widespread misperception of culture as environmentally benign, and the cultural and creative industries as being intrinsically 'clean' (Maxwell and Miller 2017, p. 177). As a result, argues Banks, there has so far been very limited environmental concern surrounding growth in the cultural and creative industries (2018, p. 370). In fact, the cultural economy is intrinsically connected to the extractive industries, both through the extraction of materials that make up the capital goods that are used for cultural production, and through the provision of energy to the sector (Maxwell, this volume). Furthermore, some sectors of the cultural and creative industries, such as media and digital industries, are highly polluting and resource intensive.

5.4 Cultural Economy, Policy and Development in Argentina

The formation of the cultural field in Argentina was influenced by the frame that was imposed in the early days of the creation of the nation state: 'Civilization or barbarism' (Svampa 1994 in Wortman 2001). This has led to a binary understanding of culture that persists until today, when some government officials speak of 'proper' culture—meaning expressions of fine arts, theatre, music and literature that follow European traditions—as distinct from and opposed to popular culture—hybrid cultural expressions emerging from the working classes, often community-based, sometimes the result of merging indigenous, afro-descendent and European cultural forms. This perception of culture is embedded in a colonial mentality, with the contemporary art and cultural production of Buenos Aires being seen as 'global' and 'trendy' and therefore highly regarded, and the cultural expressions of the provinces seen as 'backwards' (Dinardi 2015, p. 13). This implication that popular and indigenous cultures are uncivilized was and still is in fact instrumental in justifying a development model based on the expulsion and erasure of indigenous people for the appropriation of territory and the development of extractive activities.

Historically in Argentina, as in other Latin American countries, there have been structural barriers to developing the creative industries, such as unstable economies and heightened vulnerability to external financial crises (Solanas 2008). However, autonomous systems of cultural production began to emerge around the 1930s, with the growth of the middle class and the rise in rates of literacy. In Argentina, publishers and workers' libraries—mostly linked to socialist and anarchist groups—began to multiply and expand around that time. But it was only in the second half of the twentieth century that a widespread cultural market was consolidated, after the urban growth of the 1940s and the introduction of new communication technologies such as television (García Canclini 2001, p. 94–5). This was followed by an expansion of the autonomous art sphere, which brought about a prominent Avant-Garde art scene in the 1960s.

Soon after, however, came the neoliberal reform of the Argentine state, which took place in two stages: first through a military dictatorship in the 1970s, and then through a democratic government in the 1990s. Among the several transformations this entailed, was the privatization of state media and the formation of media conglomerates (Wortman 2001, p. 253). This set the stage for a hegemonic media sphere, which would be crucial to the construction of consumer identities (Ibid., p. 254). García Canclini (1987) employs the term 'neoconservative logic of privatization' to describe the approach to cultural policy adopted by countries across Latin

America as from the mid-1970s. The international economic crisis and the crisis of democracy experienced by several countries led to neoconservative currents—often embodied in military dictatorships—to push for a reorganization of the model of accumulation, the reduction of the state, de-industrialization and the monopolist concentration of production across different areas (García Canclini 1987, p. 39; Solanas 2008). To this date, ownership in the creative industries and media is still mostly concentrated in few hands, a pattern that is repeated across the region (Mastrini and Becerra 2007, p. 2). Cultural policy was subject to this process as well, and cultural production was further affected by the persecution and censorship of dissident voices. In Argentina, this depoliticisation of culture also led to a change in cultural subjectivities, from producers and participants to the aforementioned figure of consumers (Beltrán and Miguel 2012, p. 44). The neoliberal reform continued in the 1990s, with further privatization and the transfer of ownership to foreign companies, as was notably the case in the publishing industry.

Despite a profound economic and social crisis at the turn of the century, Argentina has seen an expansion of the cultural and creative industries—mostly concentrated in the city of Buenos Aires—since the early 2000s (Beltrán and Miguel 2012). This expansion process was championed by then mayor of Buenos Aires Anibal Ibarra of the centre-left alliance FREPASO, who established the Observatory for the Cultural Industries. Between 2003 and 2008, the cultural and creative industries 'constituted 8.2% of the value added of the city of Buenos Aires, with an annual growth of 14%. [...] By 2009, the creative industries represented almost 10% of the total employment in the city' (Beltrán and Miguel 2012, p. 42). In parallel to this, at the national level under the progressive governments of Nestor Kirchner and Cristina Fernández de Kirchner, Argentina saw a period of state-led democratisation of culture, with increased state investment in non-commercial cultural activities and an agenda of social inclusion.

More recently, the administrations of the city of Buenos Aires under the market oriented Pro party—from 2007 to 2015 under Mauricio Macri who would then become President, and since 2015 under Horacio Rodríguez Larreta—has championed an idea of culture as valuable for expanding tourism, and has done this through urban marketing and initiatives to attract the private sector to the cultural field, while making a significant reduction on public spending on cultural programmes. As Dinardi explains, for instance, '[t]he decision to reduce the budget of the network of community cultural centres, accompanied by eviction orders and the closure of many of these popular cultural venues, was a clear example of the GCBA's [government of the City of Buenos Aires] intention to apply a profitability logic to cultural planning' (2015, p. 14). Despite some successful cultural programmes run by the Ministry of Culture of Buenos Aires, the general approach to cultural production and cultural policy that we see under market-oriented governments can be said to be following a logic of privatization akin to that imposed in previous peaks of neoliberal rule.

Currently, the cultural and creative industries are seen as an opportunity for development and generating value (Szpilbarg and Saferstein 2014, p. 108). We can see efforts to establish the city of Buenos Aires as a global creative city through the import of international events such as Art Basel. It is telling also that the Observatory for *Cultural* Industries, which used to be part of the Ministry of Culture of the City of Buenos Aires, was replaced in 2008 by the Observatory for *Creative* Industries, which would go on to be part of the Ministry of Economic Development and be subjected to marked budget cuts for programmes aimed at developing local industries and professionalising cultural workers, despite the championing of the cultural and creative industries on a discursive level.

5.5 Cultural and Creative Industries and the Logic of Extraction

The extractive sector and the cultural and creative industries are connected in multiple ways. Although an in-depth analysis of these connections is beyond the scope of this chapter, it is important to note that the cultural and creative industries can be sites for the creation of meaning where cultural battles are fought, and are also instrumental to sustaining and legitimizing an extractivist model of development. This includes the narratives espoused by the hegemonic media (Riffo 2017), extractive companies' cultural sponsorship with the aim of obtaining a social license to operate (Serafini and Garrard 2019), and the strategic employment of environmentally-themed cultural activities as opportunities for 'greenwashing' on behalf of the state. These actions are facilitated through long-term connections between government officials, the extractive sector and the arts, and by the aforementioned concentration of ownership of the media and other cultural and creative industries.

What I am interested in exploring here, however, is how the logic of extraction—and ideas of culture, economy and development that follow this logic—are mobilized in the cultural economy. To that end, in what follows I will discuss the way in which the logic of extraction is mobilized in urban cultural dynamics, in order to show how the framework of extractivism can be useful in understanding and critiquing the expansion of the cultural and creative industries and their role in reproducing dynamics of socio-environmental inequality.

Mookerjea argues that 'space for creative economy development can only be produced through the subalternization of other modes of socio-ecological reproduction' (cited in Saifer 2018, p. 30). Such a dynamic is evident in the neighbourhood of La Boca, in the south of the City of Buenos Aires. La Boca is a neighbourhood that has traditionally been working class and home to thousands of European migrants who arrived to the country in the late nineteenth and early twentieth century. It developed as an industrial area and, following the closure of several factories in the 1970s, became a neglected part of the city (Rodríguez and Di Virgilio 2016, p. 1219), with many living in unsafe multi-family housing with no support and no planning on behalf of the state. In recent years, however, La Boca became a site of interest for the city government. In 2012 the city government issued law 4353, known as the Art District Law, and began to promote culture-led regeneration. This process was facilitated by the common combination of private and public investments and policy initiatives (Herzer and Gil y de Anso 2012, p. 14), in this case the provision of tax breaks for cultural organizations.

The regeneration of La Boca includes major infrastructure plans such as the building of a new highway connecting this southern neighbourhood of the city with the affluent northern suburbs. It also involves the construction of luxury studio flats aimed at young professionals, and the privatization of public buildings and outdoors spaces. While La Boca has had issues concerning the safety of housing for decades, *La Boca Resiste y Propone,* a local organization that works towards justice in housing, claims that the city government now speculates with old, unsafe housing. The government does not conduct the necessary renovations in order to guarantee safety in old buildings so, when fires or collapses do happen, buildings are closed down, declared uninhabitable, and later demolished, giving place to new types of housing that are unaffordable to the original inhabitants[1]. When the city government does invest in infrastructure, explains Natalia, a local activist who is part of the organization, the services they come to install have to do with 'providing guarantees for this new population they want to settle here.' The process has also been marked by a series of cultural programmes led by the city government, which display a logic of appropriation and extraction aimed at capitalizing on the cultural and historic value of the neighbourhood.

The first of these cases is *Color BA*, a mural art festival that began running in 2016, and became a point of controversy on its first edition as it brought mostly non-local artists to paint murals in a neighbourhood with a longstanding tradition of mural art. By sustaining the figure of the mural as characteristic of the area, but promoting an international 'hipster' aesthetic, the new murals, mostly painted by non-locals surrounded by a fence and guarded by the police, rapidly became the go-to image for real estate agencies selling the luxury studio flats soon-to-be open. Natalia describes this process as 'a reconfiguration of the urban space that is installed through the visual'.

The second case is that of the official sponsorship and promotion of local festivities. La Boca has a series of local festivities and traditions which include the bonfire of San Juan and the procession of the Madonna Santa. As Natalia explains, one day the city government—who had never had any involvement in organizing such festivities before—decided they would be sponsoring the celebrations. They invested in infrastructure, invited both local and high-profile artists to perform, and brought in police officers to guard the event. People who had lived in La Boca their whole lives and had always been part of organizing those festivities, suddenly received invitations to the events on behalf of the city government.

This appropriation and promotion of local art forms and traditions is of course not coincidental, and fits within an economic rationality in which cultural validation contributes to increasing property value (Zukin 1987, p. 143). What is happening in

[1] Personal interview with Natalia, Buenos Aires, 2018.

La Boca can be described as settler, culture-led and state-sponsored gentrification, and is leading to the dispossession and expulsion of its most vulnerable long-term inhabitants (Rodríguez 2015). In the Latin American context, gentrification is understood as 'an effect of neoliberal socio-spatial dynamics supported by variegated forms of symbolic and/or material displacement of low-income people, coupled with their exclusion from political decision-making' (Rodríguez and Di Virgilio 2016, p. 1218). In the same way as the frontiers of extraction are expanding in rural areas, neighbourhoods that are being gentrified can be regarded as 'new frontiers' for the expansion of the creative industries and the creative class (Saifer 2018, p. 29), in what can be understood as a process of urban extractivism. The process carries with it a form of violence, described by La Boca Resiste y Propone as institutional violence, since people's lives are being put at risk both by concrete government and private sector actions and by inaction on behalf of the same government when it comes to housing and planning.

The appropriation of local popular culture and its use for the generation of value in the real estate market follows a logic of extraction, in which culture is extracted from the local population—in some cases quite literally, as is the case of local festivities—and repackaged with the aim of serving as a tool for growth and expansion, without concern for human or environmental costs. In this sense, this case displays an example of clashing conceptions of culture and its value. While the city government and developers push for cultural regeneration as a way of increasing the value of property in the area, long-term inhabitants of La Boca uphold the popular culture, fine arts heritage and traditions of the area, their role in the community's identity and socialization, and its potential for resistance in a battle for basic human, housing and environmental rights that is taking place in the cultural sphere.

The gentrification process that is taking place in La Boca is, of course, not an isolated phenomenon. Considering this process in relation to other waves of gentrification in Latin America and at a 'planetary' scale (Lees et al. 2016) is useful for identifying patterns in the operation of transnational capital, and in the role of the creative economy as both engine and legitimising discourse of gentrification. However, the perspective of extractivism allows us to understand gentrification in relation to other sectors of the economy and to other dynamics of displacement, such as those in rural areas. Furthermore, it allows us to locate these processes symbolically within the imaginary of a country and a region that is still invested in specific visions of development based on the extraction of resources, where culture risks becoming another resource for the extraction of value only to benefit those at the top.

5.6 Towards a Post-Extractivist Cultural Sphere

Extractive development builds on the centuries-long history of Latin America as a site of extraction for the development and sustainment of capitalism and global economic powers. Similarly, the approaches to cultural policy and the cultural economy that have emerged in Argentina and other Latin American countries are the result of historical and economic processes that go back to the colonial origins of the country's cultural imaginaries. Understandings of culture shaped by modern binaries, periods of politicized culture followed by political persecution and privatization, and a national economy increasingly re-oriented towards the extractive sector are all constitutive of the way that the cultural and creative industries were set up and function today.

The perspective of extractivism allows us to draw parallels between development initiatives in rural and in urban settings, to consider how urban and rural economies and processes are imbricated, and to see how the cultural and creative industries fit within that model, developing alongside, sometimes supporting, and even enforcing similar models of extraction of value and displacement of people. In order to move towards post-extractivist futures, we must look at the sphere of cultural production to critically analyze its functioning within an extractivist model, and consider what cultural production might look like in and for a more ecologically sustainable and socially just society.

Post-extractivist visions and initiatives generally concentrate on three critical areas: generating new structures for tax revenue to replace revenues from extractive activities; developing socially relevant knowledges and technologies; and guaranteeing environmental and social justice (Martín García 2016, p. 94). If we think of a post-extractivist sphere of cultural production, the first issue to consider is what values, knowledges, processes and conceptions of culture are most compatible with the idea of a truly sustainable and democratic cultural economy. Significant changes need to occur at the root of our conceptions of culture, environment and democracy. There is a need to generate new cultural imaginaries; as Gómez-Barris asks: 'What cultural and intellectual production makes us see, hear, and intimate the land differently?' (2017, p. xx).

In this quest, it is paramount to adopt a decolonial approach to policy and practice. Such a process begins by challenging the hierarchies of value in art and culture imposed through colonialism that remain rooted to this day, and facilitating the exposure and recognition of indigenous knowledges, techniques and practices that constitute alternatives to extractive development (García Canclini 1987, p. 23). Decolonizing cultural production also means the de-centralization of the cultural and creative industries, which are mostly based in the City of Buenos Aires. It involves a shift from culture to cultures, and a non-paternalistic valuing of folklore and indigenous cultures. As I have argued earlier, in Argentina there are several cultural battles taking place at the same time, concerning both the social and economic role of cultural production, and the valuing of different cultural forms. When thinking about a post-extractive cultural sphere, it is important not to uphold one form of cultural production over another, but to allow for a multiplicity of practices based on the adoption of decolonial aesthetics and attitudes towards knowledge production and on the rejection of longstanding hierarchies of value.

A post-extractivist cultural sphere must also move away from understandings of culture that are heavily based on its economic value and be concerned with social and environmental justice. It must not see culture and nature as merely resources but rather as commons, and it must restructure the management of the commons with an emphasis on participation, democratic sustainability, and generating new ways of measuring prosperity.

Post-extractivist culture must be popular but not dominated by consumption and must generate a cultural model of active participation instead (García Canclini 1987). Indeed, parallel to the hegemonic culture of consumption, Argentina already has a strong tradition of community arts, among which embodied collective practices such as theatre, *murga*, and singing stand out (Wortman 2001, p. 255). These community and identity-based practices are embedded in non-monetary socialities and constitute a way of life, as a result embodying dissident cultural logics that coexist with the dominant one (Ibid., p. 259).

Another important issue in thinking of such radical social, economic, and cultural changes is the role of cultural policy. In a region where many local industries tend to be neglected as investment goes into infrastructure for extractive activities, should we be demanding stronger policies and investment for developing the cultural and creative industries as such? Some have pointed to the need to strengthen the cultural industries in order to contribute to diversifying the economy (Lossio Chavez 2017). This, accompanied by progressive tax reforms, could lead to a state-led socio-economic transformation in which global-facing cultural and creative industries could have an important role (Solanas 2008). Others, like Wortman, shift focus from the development of the cultural and creative industries and argue for a cultural policy that looks to strengthen the social fabric. She points to the role that cultural policy can have in rethinking social rights in the aftermath of neoliberalism. What most academics agree is that cultural policy should manage the cultural and creative industries in order to protect culture from a purely commercial logic—indeed much policy on culture at the city level in Buenos Aires is currently managed by state institutions that focus on economic development—and promote social practices that are in tune with an ethos of cultural democratization (García Canclini 1987, p. 51; Szpilbarg and Saferstein 2014, p. 106). To this we might add that cultural policy should also be in conversation with environmental policy, a key task towards developing forms of cultural production that are committed to real sustainability and preservation of local ecologies (Maxwell and Miller 2017, p. 178; Maxwell, this volume).

Finally, if we want cultural production to contribute to the building of a post-extractivist future, it is paramount to consider the connections between cultural production and other spheres of society which are working towards—or already enacting—alternatives to extractive development, such as social movements (Escobar 1992). As is the case in much of Latin America, Argentina already displays a wealth of experiments in agroecology (Palmisano 2018), horizontal organizing, cooperative production and provision of services, and solidary economics (Dinerstein 2015). Facilitating connections between cultural production and alter-development experiences in other spheres, some of which is already taking place (Merlinsky and Serafini 2019), is key to both grassroots and policy approaches to working towards a post-extractivitst society.

Acknowledgements The research conducted for this chapter was supported by a grant from The British Academy/Leverhulme Trust. I thank Lía Barrese for her insights, which have been of great value.

References

Banks, M. 2018. Creative economies of tomorrow? Limits to growth and the uncertain future. *Cultural Trends* 27 (5): 367–380.

Beltrán, G.J., and P. Miguel. 2012. Doing culture, doing business: A new entrepreneurial spirit in the Argentine creative industries. *International Journal of Cultural Studies* 17 (1): 39–54.

CEPAL. 2014. Pactos para la igualdad. http://www.cepal.org/publicaciones/xml/7/52307/2014-SES35_Pactos_para_la_igualdad.pdf. Accessed 22 Jan 2020.

Dinardi, C. 2015. Unsettling the role of culture as panacea: The politics of culture-led urban regeneration in Buenos Aires. *City, Culture and Society* 6: 9–18.

Dinerstein, A.C. 2015. *The politics of autonomy in Latin America: The art of organising hope*. Basingstoke: Palgrave Macmillan.

Escobar, A. 1992. Imagining a post-development era? Critical thought, development and social movements. *Social Text* 31 (32): 20–56.

———. 1999. *El final del salvaje. Naturaleza, cultura y política en la antropología contemporánea*. Bogotá: ICAN y CEREC.

———. 2010. Latin America at a crossroads. *Cultural Studies* 24 (1): 1–65.

Florida, R. 2002. *The rise of the creative class*. New York: Basic.

———. 2017. *The new urban crisis*. London: Oneworld.

García Canclini, N. 1987. Introducción. Políticas culturales y crisis del desarrollo: un balance latinoamericano. In *Políticas Culturales en América Latina*, ed. N. García Canclini, 13–62. México: Grijalbo.

———. 2001. *Culturas híbridas. Estrategias psara entrar y salir de la modernidad*. Barcelona: Paidós.

Gómez-Barris, M. 2017. *The extractive zone. Social ecologies and decolonial perspectives*. Durham: Duke University Press.

Gudynas, E. (2009) Diez tesis urgentes sobre el nuevo extractivismo: Contextos y demandas bajo el progresismo sudamericano actual. In Extractivismo, política y sociedad, J. Schuldt, A. Acosta, A. Barandiarán, M. Folchi, CEDLA-Bolivia, A. Bebbington, A. Alayza and E. Gudynas. Quito: CAAP and CLAES.

Hawkes, J. 2001. *The fourth pillar of sustainability. culture's essential role in public planning*. Melbourne: Cultural Development Network. http://www.culturaldevelopment.net.au/community/Downloads/HawkesJon(2001)TheFourthPillarOfSustainability.pdf. Accessed 22 Jan 2020.

Herzer, H., and L. Gil y de Anso. 2012. Introducción. In *Barrios al sur: renovación y pobreza en la ciudad de Buenos Aires*, ed. H. Herzer, 13–36. Buenos Aires: Café de las Ciudades.

Lees, L., H.B. Shin, E. López Morales, and E. 2016. *Planetary gentrification*. Cambridge: Polity Press.

Lossio Chavez, F. 2017. *From mining to videogames*. Paper presented at The Industrialization of Creativity and its Limits, St Petersburg, Russia, June 23–24.

Martín García, F. 2016. Postextractivismo y crecimiento en América Latina. Historias, problemas y desafíos hacia una agenda de luchas comunes. In *La espacialidad crítica en el pensamiento político-social latinoamericano: Nuevas gramáticas de poder, territorialidades en tensión*, ed. M. Rosales, Z. Garay Reyna, and C.E. Pedrazzani, 77–100. Buenos Aires: CLACSO.

Mastrini, G., and M. Becerra. 2007. Globalización, mercado e industrias culturales. ¿resistencia o simulacro? *Diálogos de la Comunicación* 75: 1–7.

Maxwell, R., and T. Miller. 2017. Greening cultural policy. *International Journal of Cultural Policy* 23 (2): 174–185.
Merlinsky, G., and P. Serafini. 2019. Arte y resistencias al extractivismo en Argentina. Lenguajes para defender y reinventar lo común. *Ecología Política* 57: 81–85.
Oliva Abarca, J.E. 2016. Capital cultural, creatividad y desarrollo sustentable: retos y desafíos actuales de la producción cultural. *Revista Estudios* 33: 1–24.
Palmisano, T. 2018. Alternative agricultures in the context of agribusiness. Experiences in the province of Buenos Aires, Argentina. *Estudios Sociales Revista de Alimentación Contemporánea y Desarrollo regional* 51 (28): 2–28.
Rey, G. 2009. *Industrias culturales, creatividad y desarrollo*. Madrid: Agencia Española de Cooperación Internacional para el Desarrollo.
Riffo, L. 2017. Unconventional hydrocarbons and socio-environmental conflicts. Critical analysis of the enunciative scene in the daily Río Negro about the Chevron-YPF agreement. *(En)clave Comahue* 23: 61–84.
Rodríguez, M.C. 2015. Estado, clases y gentrificación. La política urbana como campo de disputa en tres barrios de la Ciudad de Buenos Aires. In *Perspectivas del estudio de la gentrificación en México y América Latina*, ed. V. Delgadillo, I. Díaz, and L. Salinas, 205–227. Mexico: Universidad Nacional Autónoma de México.
Rodríguez, M.C., and M.M. Di Virgilio. 2016. A city for all? Public policy and resistance to gentrification in the southern neighborhoods of Buenos Aires. *Urban Geography* 37 (8): 1215–1234.
Saifer, A. 2018. Research note: Reimagining creative economy through the lens of multiple colonialisms. *Review in Cultural Theory* 8 (1): 23–38.
Serafini, P., and C. Garrard. 2019. Fossil fuel sponsorship and the contested museum: Agency, accountability and arts activism. In *Museums and activism*, ed. R.R. Janes and R. Sandell. London: Routledge.
Solanas, F. 2008. La economía creativa y las posibilidades de desarrollo en Argentina. In *Economía creativa como estrategia de desarrollo. Una visión de los países en desarrollo*, ed. A. Fonseca Reis, 161–181. Sao Paulo: Observatorio Itau Cultural.
Svampa, Maristella. 1994. El dilema argentino: Civilización o barbarie. In *De Sarmiento al revisionismo peronista*. Buenos Aires: El Cielo por Asalto.
Svampa, M., and E. Viale. 2014. *Maldesarrollo: La Argentina del extractivismo y el despojo*. Buenos Aires: Katz.
Szpilbarg, D., and E. Saferstein. 2014. De la *industria cultural* a las *industrias creativas*: un análisis de la transformación del término y sus usos contemporáneos. *Estudios de Filosofía Práctica e Historia de las Ideas* 16 (2): 99–112.
Vásquez Duplat, A.M. 2017. Presentación. In *Extractivismo urbano. Debates para una construcción colectiva de las ciudades*, ed. A.M. Vásquez Duplar, 9–14. Buenos Aires: Editorial El Colectivo.
Viale, E. 2017. Prólogo: El extractivismo urbano. In *Extractivismo urbano. Debates para una construcción colectiva de las ciudades*, ed. A.M. Vásquez Duplar, 15–20. Buenos Aires: Editorial El Colectivo.
Wortman, A. 2001. El desafío de las políticas culturales en Argentina. In *Estudios Latinoamericanos sobre cultura y transformaciones sociales en tiempos de globalización 2*, ed. D. Mato, 251–267. Buenos Aires: CLACSO.
Yúdice, G. 2002. *El recurso de la cultura. Usos de la cultura en la era global*. Barcelona: Gedisa.
Zukin, S. 1987. Gentrification: Culture and capital in the urban core. *Annual Review of Sociology* 13: 129–147.

Chapter 6
Interrogating Amazon's Sustainability Innovation

Brett Caraway

6.1 Introduction

It's October 2017 and Jeff Bezos is perched on top of a 300-foot wind turbine under a hazy blue West Texas sky. A camera-equipped drone hovers directly in front of him, revealing the rolled-up sleeves of his dress shirt, his black utility gloves, and a bottle of champagne in his right hand. He smiles slightly from under his white safety helmet and black sunglasses before smashing the bottle across the turbine. The drone pulls away, cinematically, up and to the left, giving a wider view of the surrounding windfarm. Maritime traditions of sacrificing something—like a bottle of bubbly—go back thousands of years. Christenings were done to attract favour from the gods as ship and crew ventured off into unknown and angry seas. Such ventures were part of a long and bloody history of conquest and expansion. From high up on the mast-like turbine, what storms did the diminutive Bezos see brewing on the horizon? And what part did this new windfarm play in Amazon's latest voyage of exploration and conquest?

Clearly, Amazon is making a show of their environmental programs. The company proudly advertises its sustainability initiatives which include things like packaging standards, renewable energy and energy efficiency, waste minimization, e-waste mitigation, supply chain management, technological innovation, and sustainable transportation (Amazon.com Inc 2019d). Amazon has publicly supported policies and programs designed to mitigate the impact of climate change. Since 2015, Amazon has supported the United Nations Framework Convention on Climate Change's (UNFCCC) landmark Paris Agreement, stating 'We believe that robust clean energy and climate policies can support American competitiveness, innovation, and job growth' (Amazon.com, Inc 2019c, p. 2). Additionally, Amazon has

B. Caraway (✉)
University of Toronto, Toronto, ON, Canada
e-mail: brett.caraway@utoronto.ca

K. Oakley, M. Banks (eds.), *Cultural Industries and the Environmental Crisis*,
https://doi.org/10.1007/978-3-030-49384-4_6

entered into a variety of industry partnerships ' … to accelerate the transition to a low-carbon economy' (Amazon.com, Inc 2019d).

It may seem unfair to fault Amazon for maintaining simultaneous commitments to the environment, job growth, and innovation. Nevertheless, the firm has endured criticism for its lack of transparency in accounting for its carbon footprint and environmental practices. Amazon's public commitment to providing customers with health and sustainability data has not satisfied critics. The non-profit CDP (Carbon Disclosure Project), which maintains an environmental impact disclosure database for investors, companies and citizens, recently gave Amazon an 'F' for its lack of transparency (González 2016). Similarly, Greenpeace (2017, p. 4) characterized Amazon as '… one of the least transparent companies in the world in terms of its environmental performance, as it still refuses to report the greenhouse footprint of its own operations.' Perhaps in response to these criticisms, Amazon announced in early 2019 that it would be disclosing its carbon footprint later in the year. In addition, Amazon pledged to make half of its shipments carbon-neutral by 2030 as part of its Shipment Zero initiative (Chasan 2019).

6.2 Theorizing Sustainability

Given its numerous sustainability initiatives and the recent about-face on carbon transparency, how are we to assess Amazon's role in global ecological destabilization? In this chapter, I attempt to look beneath the veneer of environmental responsibility, to ruthlessly interrogate the green rhetoric of corporate communication, recognizing the fundamental tension between economic growth and ecological sustainability. In doing so, I draw on a number of key insights from ecological economics and ecosocialist thought. I begin with the general observation that capitalism transpires in such a way as to impair its own development. In the traditional Marxian sense, this refers to the development of an antagonistic and crisis-prone set of social relations. But it also entails a crisis in the conditions of production, or the overexploitation of physical resources, resulting in higher capital costs and extensive ecological destabilization (Caraway 2017). This is not to suggest that the debate over the relationship between capitalist accumulation and ecological crisis is settled outside of ecosocialist circles. In fact, there is a widely accepted assumption that further economic growth and industrialization are required if we are to respond effectively to the challenges posed by ecological crises like climate change. From this perspective, the generation of wealth makes possible the development of technological solutions to ecological crises. Stated another way, economic growth is congruent with ecological sustainability so long as there is a transition to a low-carbon economy. The end goal of this transformation is the *circular economy*—streamlined production and economies of scale creating goods and services with zero-waste.

To critically assess the assumptions underlying this perspective, I turn to the limitations imposed on economic activity by the conservation laws of physics. The first law of thermodynamics pertains to the conservation of energy. The law tells us that while energy can exist in a variety of forms—including chemical, electrical, gravitational, heat, light, and motion—the total energy of an isolated system must remain constant regardless of any transformations in form. As Jackson (1996, p. 8) explains:

> The total energy input always matches the total energy output. For example, when coal is burned, chemical energy is transformed into thermal energy. But the heat output is no more and no less than the energy stored in the chemical bonds of the coal to start with.

A related law of conservation dictates that in an isolated system, mass is neither created nor destroyed by chemical reactions or physical transformations. That is to say, the total mass of the material inputs must be equal to the total mass of the material outputs in an isolated system. The upshot for economic activity is that '… all of the material resources which we exploit and transform through human activities must end up somewhere—if not in products, then in the environment (Jackson 1996, p. 9). Considering mass and energy together, the so-called law of conservation of mass-energy attempts to account for the fact that mass and energy can undergo conversion from one to another. Although there has been some debate about the significance of this conversion process for ecological economics (see Ayres 1999), the conservation of mass remains important to ecological economics because the energy produced or consumed during such a conversion typically accounts for an extremely small portion of the mass.

The second law of thermodynamics states that the total entropy of an isolated system increases over time. Although entropy is commonly understood as disorder, it more properly refers to the unavailability of energy in a system for conversion into work. The pertinent insight drawn from the second law of thermodynamics is that economic activity always transpires in such a way as to increase the total amount of entropy in the universe. As Georgescu-Roegen (1993b, p. 79) explains, 'matter-energy enters the economic process in a state of *low entropy* and comes out of it in a state of *high entropy*.' Georgescu-Roegen (1993a) describes how economic activity is dependent upon streams of solar energy known as *flows* and concentrated deposits of materials in the earth known as *stocks*. The fundamental limit imposed by the second law is that the terrestrial stock is subject to irreversible degradation. Hence, existing stocks are a function of their previous exploitation. Entropy ensures that economic activity is '… essentially dissipative of both energy and materials' (Jackson 1996, p. 12). Consequently, an unavoidable by-product of economic activity is increasingly disordered materials in the form of dissipative pollution arising from processes of extraction, production, distribution, consumption, and disposal. Further complicating things, flows of solar and wind energy are low intensity by comparison. As standards of living rise and economic systems becomes more complex, the demand for low entropy resources increases. And as economic development continues to irrevocably degrade finite terrestrial stocks and outstrip the capacity of global ecosystems to absorb high entropy wastes, it also diminishes the quality of life of future generations—both human and non-human.

Given these limitations, it is important to carefully consider how green corporate communication and policymaking discourses are constructed. To better understand these discourses, I will define some of the key terms. Jackson (1996, p. 2) defines *renewable resources* as those that '… are provided on a continuous basis by the flow of certain kinds of materials and energy through the environment in well-established cycles.' These include resources like oxygen, fresh water, lumber, and solar energy. The term *sustainability* is often used to assert that economic activity must bear some relation to biophysical limits. However, sustainability discourses are more often about how to manage the physical environment in order to sustain economic activity. From this perspective, the natural world exists to provide services to humanity, primarily as a resource supplier and a waste assimilator. In this context, sustainability implies two constraints: first, the exploitation of renewable resources must not exceed the regeneration rate and, second, waste flows must not exceed the assimilative capacity of the environment (Pearce and Turner 1990). Extending on this services perspective of the environment, the *circular economy* characterizes the natural world as a provider of three services: (1) a resource supplier; (2) a waste assimilator; and (3) a direct source of utility.[1] All three services are understood to have a positive economic function (that is, they can be priced and exchanged in the market) (Pearce and Turner 1990). A related, but somewhat less rigorous term, is the *closed loop economy*, typically referring to a situation in which firms recover waste flows, reuse the materials where possible, and responsibly dispose of the rest (Greenpeace 2017).

6.3 Amazon and Growth

With its corporate headquarters in Seattle, Washington, Amazon incorporated in 1994. The well-known website went live on July 16, 1995, taking in some $12,000 in orders in the first week (Stone 2013). The company went public in 1997, raising about $54 million (Stone 2013). Aside from the familiar e-commerce marketplace, Amazon also has significant operations in cloud computing, streaming content, consumer electronics, and artificial intelligence. Amazon acquired Whole Foods in 2017 and is now the second largest employer in the United States. Despite its retail operations, it is more appropriate to regard Amazon as a tech company. Amazon's value chains can be broken down into a number of component operations including product development, inbound shipping, warehousing and inventory management, e-commerce sales, payment processing, packaging, and order fulfillment (Schreiber 2016). Over the last 20 years, the company's operating strategy has increasingly centred on B2B services. In this chapter, I focus on the following aspects of Amazon's operations: logistics, cloud computing, and consumer electronics.

Although Amazon regularly reports double-digit increases in net sales, the company typically shows rather modest profits, due primarily to its considerable capital

[1] The third function refers to the aesthetic and spiritual value of the natural environment.

outlays (Khan 2017). In fact, Amazon did not report a quarterly profit until the end of 2001. From the beginning, Amazon chose to 'prioritize growth' over short-term profits because Bezos believed that 'scale is central' to Amazon's business model (Amazon.com, Inc 1998, p. 6). The initial decision to prioritize growth has paid off for the firm. Amazon's 2017 Annual Report indicates that sales increased 27% and 31% in 2016 and 2017 (Amazon.com, Inc 2018a). Amazon's 2018 fourth quarter sales continued their skyward trajectory, increasing 20% to $72.4 billion. Net sales for 2018 increased 31% to $232.9 billion, compared with $177.9 billion in 2017 (Amazon.com, Inc 2019a). The emphasis on growth and scale remains a consistent part of Amazon's operational strategy: 'We are rapidly and significantly expanding our global operations, including increasing our product and service offerings, and scaling our infrastructure to support our retail and services businesses' (Amazon.com, Inc 2018a, p. 6). Amazon's profits are also a function of other company's sales as it offers a variety of e-commerce and B2B services including storage, database, computation, fulfillment, publishing, digital content, advertising, and co-branded credit cards. In 2017, more than half of the units sold worldwide on Amazon were from third party vendors (Amazon.com, Inc 2018a, p. 21).

Amazon.com, Inc (2018a, p. 6) thus views any potential decrease in demand as a form of risk in that it would impede growth:

> Our revenue and operating profit growth depends on the continued growth of demand for the products and services offered by us or our sellers, and our business is affected by general economic and business conditions worldwide. A softening of demand, whether caused by changes in consumer preferences or a weakening of the U.S. or global economies, may result in decreased revenue or growth.

Bezos (Amazon.com, Inc 2018a) speaks of the demand that fuels economic growth in a distinctly personal tone:

> One thing I love about customers is that they are divinely discontent. Their expectations are never static—they go up. It's human nature. We didn't ascend from our hunter-gatherer days by being satisfied. People have a voracious appetite for a better way, and yesterday's 'wow' quickly becomes today's 'ordinary.' (p. 3)

Clearly, Bezos delights in meeting consumer demand—perhaps, even driving it. Equally clear then, Amazon's prioritization of growth must inform any assessment of its sustainability initiatives.

Amazon advertises their global teams that focus on environmental issues and innovations in sustainability. Amazon boasts of programs directed at packaging, renewable energy and energy efficiency, waste diversion, e-waste management, supply-chain management, sustainable transportation, and technological innovation. Amazon's Sustainability Question Bank states:

> Our Circular Economy team conducts surveys of our fulfillment centers to understand the items that make up our waste stream. We use the data from these surveys to evaluate existing recycle processes, analyze new diversion and reuse opportunities, and engage multiple business units around how to improve waste diversion. (Amazon.com, Inc 2019d)

In the same document, Amazon proudly describes how their Circular Economy team donates products to charitable organizations—which has the added benefit of increasing the rate of inventory turnover for the company. Amazon describes its approach to the Circular Economy as consisting of the following initiatives: (1) packaging; (2) curbside recycling; (3) waste diversion; (4) e-waste management; and (5) used product resale (Ibid.). Amazon also advertises its membership in a number of trade organizations including Business for Social Responsibility, the Closed Loop Fund, The Recycling Partnership, Sustainable Fuel Buyer's Principles, and the Sustainable Packaging Coalition among others. Generally speaking, these consortia promote private-sector solutions to ecological crises, based chiefly on the extension of private property rights, voluntary compliance, supply chain integration, and technological innovation. In order to assess Amazon's various sustainability initiatives, I offer the following analyses of Amazon's logistics, cloud computing, and consumer electronics operations.

6.4 Logistics

One of Amazon's truly remarkable feats has been the creation of a tightly controlled, massive logistics system. To accommodate growth, the firm quickly transitioned from batch orders to an infrastructure enabling uninterrupted and predictable inventory turnover (Stone 2013). The firm initiated a series of investments in labour productivity, facilities, and inventory management software. As Stone (2013, p. 173) explains:

> Tightly controlling distribution allowed the company to make specific promises to customers on when they could expect their purchases to arrive. Amazon's operating all of its own technology, from the supply chain to the website, allowed [Amazon] to create algorithms that modeled countless scenarios for each order so systems could pick the one that would yield the quickest and cheapest delivery. Millions of those decisions could be made every hour, helping Amazon reduce its costs—and thus lower prices and increase volume of sales.

Amazon's investment in logistics is ongoing. Its operations are becoming increasingly complex as it expands its fulfillment and data centre capacity. Amazon makes these logistics services available to both customers and other businesses. The efficiency of its fulfillment network and data centres allows Amazon to optimize inventories, minimize shipping costs, and forecast consumer demand with surprising accuracy. Amazon's shipping costs in 2017, which include sortation, delivery centre, and transportation costs (inbound shipping costs to Amazon are included in final selling price) were $21.7 billion, a 34% increase from the previous year (Amazon.com, Inc 2018a, p. 27). As Amazon adds additional services, more expensive shipping options, and lower shipping rates, these costs are expected to grow.

In 2000, Amazon started offering holiday-season free shipping to customers who placed orders of $100 or more. Amazon made the promotion permanent in 2002, lowering the minimum order to $25, and setting the stage for the introduction of Amazon Prime the following year (Stone 2013). Prime is a paid subscription service, giving members access to a variety of premium services, including free two-day delivery. Amazon boasted 20 million Prime subscribers in 2016. In 2017, it shipped more than 5 billion packages worldwide through Prime (Carman 2018). By 2018, Prime membership exceeded 100 million globally (Amazon.com, Inc 2018a). Amazon is committed to making Prime shipping faster and expanding the availability of Prime FREE Same-Day Delivery and Prime FREE One-Day Shipping programs. Amazon.com, Inc (2018d) also announced that it plans to bring the premium service to more cities by leasing an additional 10 Boeing 767-300 cargo aircraft and adding 4 gateway operations to the onsite facilities it already operates at over 20 airports.

Concurrently, Amazon.com, Inc (2019d) asserts that it is '... committed to minimizing our carbon emission by optimizing our transportation network, improving product packaging to drive efficiency in the distribution of products.' The company is a signatory to the Sustainable Fuel Buyers' Principles, a business consortium directed at achieving greater efficiencies in road freight transportation systems. Amazon.com, Inc (2018c) boasts of the fuel efficiency of its North American fleet of semi-trailers, equipped with aerodynamic skirts and PSI automatic tire inflation systems, and of the upcoming launch of its first low-pollution last-mile fleet of electric and natural gas delivery vehicles in Europe.

Nevertheless, the ecological impact of Amazon's logistics infrastructure has come under scrutiny. Preliminary research comparing the carbon footprints of online shopping with brick and mortar retail seemed to suggest that e-commerce was a more environmentally responsible alternative (Weber et al. 2009; Weideli 2013). Researchers believed that the more efficient delivery systems of parcel services were preferable to shoppers driving to brick-and-mortar stores (Chung 2018). Other research has cast doubt on these findings however. Wygonik and Goodchild (2012) found that the largest savings in CO_2 emissions were associated with delivery systems serving low population densities, though the profitability of delivery systems is dependent on high population densities. These same authors also found that while last-mile passenger travel resulted in the highest amount of vehicle miles traveled, the efficiency of delivery services did not offset the higher pollution of delivery trucks (Wygonik and Goodchild 2014). Laghaei et al. (2016) studied the impact of online shopping on the transportation infrastructure of Newark, Delaware. Based on the projected volume of purchases, they found that online shopping placed additional burdens on the transportation network as measured by travel time, delay, speed, and pollution. Similarly, the environmental organization 350 Seattle (2018) estimated that in 2017, Amazon's shipping operations released at least 19.1 million metric tons of CO_2 into the environment, the equivalent of about 100,000 rail cars worth of burned coal (Coombs 2018). Amazon has an economic incentive to offer expedited shipping as one of their competitive offerings, even though the practice is less efficient than consolidated orders paired with slower shipping options. And where Amazon does achieve efficiencies, the growth and expansion of its operations is likely to wipe out any resultant ecological benefits.

Nevertheless, Amazon pursues a number of initiatives to mitigate the environmental impacts of its logistics system. Amazon features a package recycling guide on their website. It attempts to use 100% recyclable packaging and to ship products in their own packages whenever possible. It seeks to reduce packing materials and shipping supplies, thereby reducing costs. Amazon.com, Inc (2019d) asserts:

> Over the past 10 years, our sustainable packaging initiatives have eliminated more than 244,000 tons of packaging materials, avoiding 500 million shipping boxes. Our programs have reduced packaging waste by 16%, avoiding 305 million shipping boxes in 2017 alone.

Amazon is party to a number of industry consortia related to packaging and recycling programs. It is a member of the Sustainable Packaging Coalition, an industry association pursuing greater efficiencies in packaging systems to align 'economic prosperity and a sustainable flow of materials' (Amazon.com, Inc 2019c). Amazon is also part of The Recycling Partnership, a US non-profit that provides support to local communities and governments to encourage the expansion of curbside recycling programs. Additionally, Amazon committed $10 million to the Closed Loop Fund, a social investment program providing low interest loans to private companies and municipalities to develop waste diversion infrastructure. The investment is intended to bring curbside recycling to 3 million homes across the US. The Closed Loop Fund aims to '… eliminate more than 16 million tons of greenhouse gas [and] divert more than 8 million cumulative tons of waste from landfills' by 2028 (Amazon.com Inc 2018b, ¶ 4).

Amazon's use of words like 'eliminated', 'avoiding', and 'reduced' require careful attention. In everyday usage, these terms might indicate that something was either prevented or bypassed in an enduring fashion, and that the resultant savings exist in perpetuity. However, this is not how the terms are being used here. Amazon simply means that it has consumed less of a given stock of resources in the present moment in order to consume more of the stock in the future. Simply put, Amazon is deferring its consumption of resources until some later point in time. Even the name Closed Loop Fund is misleading in that it seems to suggest that there is an economic model in which waste never leaves the system. Such a model is thermodynamically impossible. As Jackson (1996, p. 55) explains:

> The only way we could reverse the dissipation of materials from human activities would be by using vast amounts of energy which in their turn could only be obtained from material resources. We could come close to closing a particular material cycle if we supplied sufficient high-quality energy to the task. But supplying this energy is itself a thermodynamic process, unavoidably dissipating more materials and more energy.

6.5 Amazon Web Services (AWS)

Amazon Web Services (AWS) is a subsidiary of Amazon supplying businesses with on-demand cloud services like networking, storage, databases, analytics, developer tools, mobile, and computation. AWS clients include high profile firms and organizations like Pinterest, Instagram, Netflix, NASA, and the CIA (Stone 2013). Since its launch in 2002, AWS has evolved into a global array of server farms providing clients with large scale computing capacity on a per-usage basis. Amazon leverages its scale to offer these services at low prices. AWS has been integral to the success, innovation, and growth of countless firms and organizations. It also transformed Amazon into something more than an online retail firm. As a portion of Amazon's total operating income, AWS grew from 38 percent in the first quarter of 2015 to 52 percent in the third quarter of 2015 (Galloway 2017). As of 2017, it boasts a run rate of $20 billion (Amazon.com, Inc 2018a). Amazon continues to expand AWS with services like machine learning and artificial intelligence, Internet of Things, and serverless computing.

'It's greener in the cloud' proclaims Amazon from the top of its *AWS and Sustainability* webpage (Amazon.com, Inc n.d.). The company has committed to 100% renewable energy for its AWS operations in the long term. Amazon.com, Inc (2019d) was the leading corporate purchaser of renewable energy in the US in 2016. The Amazon Wind Farm Texas, mentioned at the beginning of this chapter, is just one part of a global infrastructure that includes 53 wind and solar projects. The company estimates that its investments in renewable energy will one day generate 3.6 million MWhours of energy annually (Amazon.com, Inc 2018c). Amazon boldly asserts that using its cloud services is inherently environmentally beneficial:

> A typical large-scale cloud provider achieves approximately 65% server utilization rates versus 15% on-premises, which means when companies move to the cloud, they typically provision fewer than ¼ of the servers than they would on-premises. In addition, a typical on-premises data centre is 29% less efficient in their use of power compared to a typical large-scale cloud provider that uses world-class facility designs, cooling systems, and workload-optimized equipment. Adding these together (fewer servers used plus more power efficient servers), customers only need 16% of the power as compared to on-premises infrastructure. This represents an 84% reduction in the amount of power required. (Amazon.com, Inc n.d. p. 2)

The underlying message of the 'environmental benefits inherently associated with running applications in the cloud' (Amazon.com, Inc n.d.) is that leveraging Amazon's scale is green: '… our scale allows us to achieve higher resource utilization and energy efficiency' (Amazon.com, Inc 2019d).[2] Yet the efficiency gains associated with scale are not likely to offset the levels of resource depletion required to achieve scale in the first place, let alone maintain or grow it. According to Pearce (2018, p. 6), data centres now use up more than 2% of the world's electricity and release an equivalent amount of CO_2 as the airline industry:

> Storing, moving, processing, and analyzing data all require energy. Lots of it. The processors in the biggest data centers hum with as much energy as can be delivered by a large power station, 1,000 megawatts or more. And it can take as much energy again to keep the servers and surrounding buildings from overheating.

While Amazon states that it is pursuing greater energy efficiency by selecting appropriate water efficient cooling methods, many of Amazon's data centres are located in places like São Paulo, Mumbai and Singapore with tropical climates where more energy is required to prevent overheating. This is but one example of the paradoxical

[2] In this context *higher resource utilization* refers to Amazon's scalable infrastructure.

nature of sustainability initiatives premised on leveraging economies of scale. And while Amazon's long-term commitment to 100% renewable energy for AWS may shift the relative share of the firm's energy consumption from non-renewable to renewable resources, there is no reason to believe this will result in a net reduction in the consumption of non-renewable resources. Given that one of the primary benefits of AWS is its capacity to drive economic growth by allowing small to medium sized enterprises to leverage Amazon's massive economies of scale, it seems probable that AWS contributes to, rather than reduces, the accelerating depletion of non-renewable resources.

6.6 Amazon Consumer Electronics

Amazon manufactures a number of popular consumer electronics including the digital personal assistant Alexa, the smart speaker system Echo, the Kindle e-reader, and the digital media player Fire TV among others. Sales of Amazon consumer electronics have grown enormously over the years. For example, Amazon has sold in excess of 100 million Alexa devices (Bohn 2019) and 2017 was Amazon's best year to-date in hardware sales (Amazon.com, Inc 2018a). Amazon's success with its consumer electronics has come as the electronics industry itself has weathered increasing criticism for its environmental impacts. Greenpeace (2017) observes that while many devices are now more energy-efficient, they are also more complex, increasing the amount of resources necessary for their manufacture. It is difficult to estimate the greenhouse emissions arising from the processes of resource extraction, manufacture, distribution, usage, and disposal, because firms like Amazon typically only report information about their own operations, and not their suppliers. So, even if Amazon carries through on its promise to provide information about its carbon footprint, we are unlikely to learn much about the total impact of its supply chain.

Amazon maintains that its manufacturing supply chain meets high standards of ethical and legal expectations. Amazon's Supplier Code of Conduct focuses primarily on issues of health and safety, wages and benefits, working conditions, and non-discriminatory policy (Amazon.com, Inc 2019b). Its supply chain standards prohibit conflict minerals and require that suppliers comply with applicable environmental laws. Amazon.com, Inc (2019d) has a Chemicals Policy regulating mutagens, carcinogens, and other toxic substances. The company offers a recycling program for discarded electronics, including Amazon devices and select accessories. Amazon also offers gift cards to customers who want to trade in old electronic devices. The company has a promotion in which customers are able to exchange select electronic devices for credit towards new Amazon products like Fire Tablet, Echo, Kindle, and Fire TV. In addition, Amazon created an online store, called Amazon Warehouse, which offers returned items for resale. Amazon even covers shipping costs to one of their certified recyclers for devices that are not eligible for trade-in.

On their face, these programs appear admirable. Yet there is a thermodynamic paradox in Amazon's recycling initiatives. To begin with, entropy places unavoidable limits on the recycling of any material. Furthermore, any attempt to close a material loop involves the expenditure of even greater levels of energy. In other words, Amazon's returned items and recycled goods don't just magically appear at a certified recycler or Amazon Warehouse—these recycled items enter into a global logistics system with its own energy requirements. Electronics recycling necessarily involves additional energy consumption and dissipative pollution to feed a vertical supply chain consisting of inbound shipping, product refurbishment, packaging, retailing, and distribution. Worse yet, these programs may actually encourage overconsumption and unnecessary returns as consumers begin to underestimate the environmental impact of their own actions. Moreover, it is imperative to recognize that these waste diversion programs are fundamentally compromised by Amazon's grander commitment to increasing the volume of its sales. Waste diversion programs are attractive to firms like Amazon only to the extent that they are aligned with that objective. Trade-in programs can stimulate demand for new products, incentivizing both product obsolescence and rapid inventory turnover. Far removed from any appreciable contribution to ecological stabilization, Amazon's waste diversion initiatives are more likely designed to sustain the company's growth.

6.7 The Problem of Scale

In September 2017, the Seattle Times was interviewing Amazon employees about the possibility of the company establishing a second headquarters in another North American city. AWS software development engineer Chad Miller commented: 'I'm happy to distribute our people to places that can sustain us … our growth here is unsustainable' (Lee 2017, p. 6). The comment is instructive in that it reflects an unmistakable corporate ethos. For Amazon, sustainability is about sustaining Amazon. It is not about sustaining global ecologies. In its communication with investors, Amazon is remarkably clear about how it views its relationship to ecological and environmental issues. When assessing the risk posed by suppliers who might implement hazardous environment practices, Amazon.com, Inc (2018a, p. 13) only observes that '… it could damage our reputation, limit our growth, and negatively impact our operating results.'

All crises, including ecological crises, are disciplinary in that they compel capital to restructure its processes of exploitation and accumulation (O'Connor 1988). What the foregoing analysis makes clear is that Amazon is using this occasion to advance economic growth as the solution to ecological destabilization. As Amazon's output rises, the average cost of producing its goods and services falls. The resultant savings, we are told, allows Amazon to look for greater efficiencies and to pursue sustainability innovations. Scale allows Amazon to achieve higher resource utilization in its data centres, to use its purchasing power to create more opportunities for renewable energy, and to leverage its AWS infrastructure in support of ecological research. Amazon.com, Inc (2019c) states, 'We remain committed to putting our scale and inventive culture to work in ways that are good for the environment and our customers.' Economic growth becomes the solution to an ecological crisis which is itself an outcome of expanding economic activity. And while market competition and profit incentive have been a driving force of innovation and efficiency, these dynamics are in the service of capital—not the natural world. They drive the material throughput which imperils the future.

At best, Amazon's sustainability initiatives fall far short of reversing the entropic dissipation of resources; at worst, they dramatically increase the rates of material throughput by encouraging profligate consumption. There is a fundamental incompatibility between ecological sustainability and Amazon's sustainability. Like many large firms, Amazon's pursuit of lower costs means that it does more with less people (Galloway 2017). Yet Amazon's power of scale comes with its own set of potential crises, for which the only palliative option appears to be more economic growth. Amazon, like all firms, is subject to the coercive dynamics of competition, and must bring the world into accord with its own requirements. Unfortunately, there is no necessary correspondence between those necessities and the sustainability of the natural world.

References

350 Seattle. 2018. Amazon. *350 Seattle*. http://350seattle.org/amazon/. Accessed 30 Mar 2019.

Amazon.com, Inc. 1998. 1998 Letter to shareholders. Amazon investor relations. https://ir.aboutamazon.com/static-files/589ab7fe-9362-4823-a8e5-901f6d3a0f00. Accessed 30 Mar 2019.

———. 2018a. 2017 Amazon annual report. Amazon investor relations. https://ir.aboutamazon.com/static-files/917130c5-e6bf-4790-a7bc-cc43ac7fb30a. Accessed 30 Mar 2019.

———. 2018b. Amazon Invests $10 Million in Closed Loop Fund to increase recycling rates across the country. *BusinessWire*, October 16. https://www.businesswire.com/news/home/20181015006009/en/Amazon-Invests-10-Million-Closed-Loop-Fund. Accessed 23 Jan 2020.

———. 2018c. *Energy and Environment*. https://www.aboutamazon.com/sustainability/energy-and-environment/energy-and-environment. Accessed 30 Mar 2019.

———. 2018d. *Press release: Amazon's air network expands to support the growth of Prime fast, free shipping for customers*. Amazon. https://press.aboutamazon.com/news-releases/news-release-details/amazons-air-network-expands-support-growth-prime-fast-free. Accessed 30 Mar 2019.

———. 2019a. *Amazon.com announces fourth quarter sales up 20% to $72.4 billion*. Amazon investor relations. https://ir.aboutamazon.com/news-releases/news-release-details/amazon-com-announces-fourth-quarter-sales-20-724-billion. Accessed 23 Jan 2020.

———. 2019b. *Responsible Sourcing*. Amazon. https://www.aboutamazon.com/sustainability/responsible-sourcing. Accessed 30 Mar 2019.

———. 2019c. *Sustainability Partnerships*. Amazon. https://www.aboutamazon.com/sustainability/sustainability-partnerships. Accessed 23 Jan 2020.

———. 2019d. Sustainability question bank. Amazon. https://www.amazon.com/qb. Accessed 23 Jan 2020.

———. n.d. *AWS & sustainability*. Amazon. https://aws.amazon.com/about-aws/sustainability/. Accessed 30 Mar 2019.

Ayres, R.U. 1999. The second law, the fourth law, recycling and limits to growth. *Ecological Economics* 29 (3): 473–483.

Bohn, D. 2019. Amazon says 100 million Alexa devices have been sold—What's Next? *The Verge*, January 4. https://www.theverge.com/2019/1/4/18168565/amazon-alexa-devices-how-many-sold-number-100-million-dave-limp. Accessed 30 Mar 2019.

Caraway, B. 2017. Literal media ecology: Crisis in the conditions of production. *Television and New Media* 19 (5): 486–503. https://doi.org/10.1177/1527476417712459.

Carman, A. 2018. Amazon shipped over 5 billion items worldwide through Prime in 2017. *The Verge*, January 2. https://www.theverge.com/2018/1/2/16841786/amazon-prime-2017-users-ship-five-billion. Accessed 30 Mar 2019.

Chasan, E. 2019. Amazon's new environmental report will show how bad 2-day shipping is. *Bloomberg*, February 20. https://www.bnnbloomberg.ca/amazon-s-new-environmental-report-will-show-how-bad-two-day-shipping-is-1.1217273. Accessed 30 Mar 2019.

Chung, E. 2018. Want it tomorrow? Some online shopping habits are terrible for the environment. *CBC*, November 21. https://www.cbc.ca/news/technology/online-shopping-carbon-footprint-1.4914942. Accessed 30 Mar 2019.

Coombs, C. 2018. Seeing Amazon shareholder meeting as a platform, group says e-commerce giant hiding carbon footprint. *Puget Sound Business Journal*, May 30. https://www.bizjournals.com/seattle/news/2018/05/30/amazon-shareholder-meeting-carbon-footprint.html. Accessed 30 Mar 2019.

Galloway, S. 2017. *The four: The hidden DNA of Amazon, Apple, Facebook, and Google*. New York. Portfolio/Penguin.

Georgescu-Roegen, N. 1993a. Selections from 'Energy and Economic Myths'. In *Valuing the earth: Economics, ecology, ethics*, ed. H.E. Daly and K.N. Townsend, 89–112. Cambridge: MIT Press.

———. 1993b. The entropy law and the economic problem. In *Valuing the earth: Economics, ecology, ethics*, ed. H.E. Daly and K.N. Townsend, 75–88. Cambridge, MA: MIT Press.

González, Á. 2016. Amazon gets an 'F' from the carbon disclosure project. *Seattle Times*, November 1. https://www.seattletimes.com/business/amazon/amazon-reluctant-to-share-carbon-emissions-data/. Accessed 30 Mar 2019.

Greenpeace. 2017. *Guide to greener electronics, 2017*. Washington, DC: Greenpeace Reports.

Jackson, T. 1996. *Material concerns: Pollution, profit and quality of life*. London: Routledge.

Khan, L.M. 2017. Amazon's antitrust paradox. *The Yale Law Journal* 126: 710–805.

Laghaei, J., A. Faghri, and M. Li. 2016. Impacts of home shopping on vehicle operations and greenhouse gas emissions: multi-year regional study. *International Journal of Sustainable Development and World Ecology* 23 (5): 381–391.

Lee, J. 2017. 'Seattle will always be home': Amazon employees, others react to news of tech giant's second HQ. *Seattle Times*, September 7. https://www.seattletimes.com/business/amazon/seattle-will-always-be-home-amazon-employees-others-react-to-news-of-tech-giants-second-hq/. Accessed 30 Mar 2019.

O'Connor, J. 1988. Capitalism, nature, socialism: A theoretical introduction. *Capitalism, Nature, Socialism* 1 (1): 11–38.

Pearce, D.W., and R.K. Turner. 1990. *Economics of natural resources and the environment*. Baltimore: The Johns Hopkins University Press.

Pearce, F. 2018. Energy hogs: Can world's huge data centers be made more efficient? *Yale Environment 360*, April 3. https://e360.yale.edu/features/energy-hogs-can-huge-data-centers-be-made-more-efficient. Accessed 30 Mar 2019.

Schreiber, Z. 2016. Is logistics about to get Amazon'ed? *Techcrunch*, January 29, https://techcrunch.com/2016/01/29/is-logistics-about-to-get-amazoned/. Accessed 30 Mar 2019.

Stone, B. 2013. *The everything store: Jeff Bezos and the age of Amazon*. New York: Back Bay Books.

Weber, C.L., C.T. Hendrickson, H.S. Matthews, A. Nagengast, R. Nealer, and P. Jaramillo. 2009. Life cycle comparison of traditional retail and e-commerce logistics for electronic products: A case study of buy.com. IEEE International Symposium on Sustainable Systems and Technology. IEEE.

Weideli, D. 2013. *Environmental analysis of US online shopping*. Cambridge: MIT Center for Transportation and Logistics.

Wygonik, E., and A. Goodchild. 2012. Evaluating the efficacy of shared-use vehicles for reducing greenhouse gas emissions: A U.S. case study of grocery delivery. *Journal of the Transportation Research Forum* 51 (2): 111–126.

———. 2014. Comparison of vehicle miles traveled and pollution from three goods movement strategies. *Sustainable Logistics* 6: 63–82.

Chapter 7
Re-thinking the Creative Economy Through Informality and Social Inclusion: Changing Policy Directions from Latin America

Cecilia Dinardi

7.1 Introduction: The Promise and Failure of the Creative Economy

In recent decades, the creative economy, comprising the so-called cultural and creative industries, has risen as the new global orthodoxy (Schlesinger 2017) with the potential to revive declining cities and boost growth in post-industrial contexts. Across the globe, new incubation schemes, observatories, districts and tax-reduction initiatives have been created to support the new creative economy at municipal, state and national levels. The allure of creative industries, particularly their promise for urban regeneration, city branding and tourism, has become institutionalized amongst new governmental agencies, industry organizations and support schemes. Originally constructed as an area of policy intervention in the global North, the promise of the creative economy was defined by its bohemian, entrepreneurial, open and meritocratic image, according to Richard Florida, where anyone could in principle benefit from this new economy. The UK appeared at the forefront of the academic debates and policy developments in this area; the work of Charles Landry (2000) originally encouraged the use of creativity as a tool for social inclusion and innovation in cities.

However, evidence of the existence of racial, class and gender inequalities within and between the cultural and the creative industries is ever-growing (Conor et al. 2015; Gill 2014; Hesmondhalgh and Baker 2011; O'Brien and Oakley 2015; Saha 2017). There is also a need to shed light on the *informal* creative economy that develops outside the radar of formal measurements and public regulation and support. The informal creative economy is diverse by definition—including for profit and non-profit cultural activities, public and private actors, individuals and

C. Dinardi (✉)
Institute for Creative and Cultural Entrepreneurship, Goldsmiths, University of London, London, UK
e-mail: c.dinardi@gold.ac.uk

K. Oakley, M. Banks (eds.), *Cultural Industries and the Environmental Crisis*,
https://doi.org/10.1007/978-3-030-49384-4_7

organizations—and is interdependent and intertwined with the formal economy, making it difficult to draw clear boundaries between the two. As such, it is outside the reach of most public policies. Over 60% of the world's population work in the informal economy. The majority are in emerging and developing countries and lack decent working conditions, social protection and labour rights (International Labour Organization 2018). Examples within the informal creative economy include non-taxed or monitored activities that are not covered by formal arrangements, such as busking and street performances, community radio stations, small, underground music venues or squatted arts spaces, amongst others. There is striking evidence of the precariousness, exploitation, work insecurity and low (or no) paying jobs that affect those working in the cultural and creative industries (Banks 2007; Curtin and Sanson 2016; Gill and Pratt 2008; Merkel 2018; O'Brien et al. 2016). Despite the existing myth of meritocracy, these industries benefit the already privileged in society (Brook et al. 2018; Littler 2017). Research conducted in this area has equally shown the need to develop situated and historical analyses of the conditions of production in the creative sector (Banks et al. 2013), and this includes the production that takes place in peripheral areas such as informal settlements. Yet such rich, dispersed and economically and socially significant cultural production has been largely overlooked in dominant policy and academic accounts.

In Latin America, the creative economy has been a rapidly emerging area of cultural, economic and urban policy development. The work of international organisations such as UNESCO and UNCTAD, particularly their Creative Economy Report Series (2008, 2010) with follow-up special editions and outlooks (2013 and 2018), has been pivotal in encouraging the creation of new institutional frameworks in support of the cultural and creative industries, especially in developing countries. Equally, municipal and national governments have been investing in the construction of new cultural indicators with the hope of demonstrating the economic contributions of the creative sector. Despite recent publications on the creative economy in Latin America and the Caribbean—for example, the Orange Economy commissioned by the Inter-American Development Bank in 2013—existing data is not sufficient, accurate, up-to-date or comprehensive enough to provide evidence to support the design of new creative economy policies in the region. Nonetheless, there are a range of freshly-designed initiatives: cultural cards containing monthly public credit for young people in high school to spend on cultural consumption (for example, cinema, theatre, museums, concerts, books, etc.) in Uruguay, Brazil and Argentina; public-private sector coalitions in Colombia; digital platforms in Mexico; specialized markets and districts in Argentina; technology hubs in Peru; creative economy forums in Nicaragua; and creative industry mapping documents in Chile, among many others. Despite these new developments, the contribution of the creative sector to employment and national economies remains largely invisible to both the general public and the official government measurements (Buitrago and Duque 2013).

This chapter examines whether a focus on the informal creative economy can support peripheral cultural scenes that remain invisible to policy and society. The first part briefly outlines key features of informal settlements in Latin America. The second looks at policy initiatives in Rio de Janeiro and Buenos Aires that have sought to develop, encourage and support creative economies in disadvantaged areas. The third part reflects how these initiatives might help expand, challenge or rethink predominant views and policies on the creative economy, assessing whether they might also help overcome 'creative injustice' in the sector (Banks 2017). Overall, the chapter shows how Latin American countries are starting to launch initiatives for the creative sector that are motivated by concerns with social inclusion and development in marginalized locations. However, these goals can be at odds with the frameworks from which the programmes develop, as they are often underpinned by a market-oriented logic based on individualistic entrepreneurialism and commercialisation.

7.2 Creative Economies in Disadvantaged and Peripheral Areas

With a strikingly large concentration of informal settlements in the cities of the global South, Latin America and the Caribbean remain amongst the most urbanized and unequal regions on the planet (UN-Habitat 2012). This shows both the resilience of urban dwellers, expressed in their self-made constructions and community-led undertakings, and the major deficit that exists in public policy provision of decent housing and public services which has historically been insufficient, failing to address the basic needs of the poorest sectors of the population.

In Latin America and the Caribbean one third of the population lives in slums. These informal houses are more prone to suffer climate-related natural disasters, as housing is one of the most affected sectors in the aftermath of earthquakes, floods, fires and landslides (Palma and Garcia Lozano 2017). Informal settlements are usually stigmatized as spaces of violence, poverty and crime, and characterized through their precariousness and deprivation in relation to existing formal urban patterns (De Souza e Silva 2014, p. 61). Yet they have also been embraced—and, to some extent, romanticized—for their resilient, sustainable and entrepreneurial spirit that generates innovation, creativity and social experimentation, making 'permanent' that which is 'in process' (Jauregui 2011). Lessons can be drawn from these locations in terms of sustainable urban design, self-made housing constructions, collaborative undertakings, energy efficiency, recycling mechanisms and resilience and adaptation. In other words, this evokes an image of the 'city as a high-tech favela':

> An urban zone that avoids rigid solutions and fosters reversible structures that can be dismantled and transformed, accommodating new, unforeseen activities. Imagine a perpetually unfinished city. Buildings are mixed-use: workshops, light industry and agriculture are interlaced with residences, sports fields, schools and markets … The city is alive 24 hours a day. If this seems like the ultimate nightmare, we see a potential mode of sustainability and tolerance. (Brillembourg et al. 2015, p. 102)

Equally, a different model of urban creativity is emerging from these harsh contexts where young artists and cultural producers, together with local NGOs and co-operatives, are organizing and participating in cultural activities that challenge common stereotypes that associate them with poverty, violence and criminality. In the context of urban extremes, these creative spaces of cultural production contribute to the vitality of informal urbanisms and, in so doing, unsettle dominant representations that depict them as sites of infrastructural poverty and social exclusion (Mbaye and Dinardi 2019). Community-led and self-managed, these activities range from orchestras, ballet, break dance, circus and photography to hip-hop, ceramics, cinema and theatre, amongst many others. As part of a wider enchantment with the creative economy, local governments have launched programmes in support of artistic and cultural activities specifically targeted at shanty towns and slums. This Chapter analyzes two ongoing examples, from Rio de Janeiro and Buenos Aires.

If the future resilient city is in the global South (Brillembourg et al. 2015), the prospects of sustainable creative economies might well be located in the informal city, in the cultural and artistic production that takes place in slums, shanty-towns and impoverished neighbourhoods, based on networks of collaboration, solidarity economies, adaptation, and inventiveness in the use of limited resources. Urban informality involves a relational process of cultural, urban and political contestation that can affect the whole city. This is being constituted by kinetic, in motion, temporary, spontaneous, flexible, recycled and reinvented spaces (Mehrotra 2010), 'a new way of life' that spreads beyond the poor or the marginalized (Alsayyad 2004). This poses a major challenge for urban cultural policy—how can these informal economies of creativity be supported by novel outlooks? How can we avoid subjecting them to formalization and entrepreneurship schemes that are oriented towards commercializing individual products?

7.3 Favela Criativa, Rio de Janeiro

In Brazil, the creative economy has been rapidly developing, experiencing continuous growth and investment despite the recent severe institutional and economic crises. A stepping stone in the institutional consolidation of the field was the creation of the Secretariat of Creative Economy (SEC) in 2011 as part of the Ministry of Culture, and the launch of several editions of official documents mapping the performance of the creative industries and measuring their economic value (FIRJAN 2008, with updates every 2 years). The latest mapping document (FIRJAN 2019) shows some recovery from these crises, with a stable participation of creative industry workers in the formal labour market, workers who receive salaries 2.45 times higher than the average for the rest of the economy. 50% of the creative jobs in the country are concentrated in São Paulo and Rio de Janeiro. The state of Rio de Janeiro has been at the forefront of producing innovative developmental policies in support of the creative economy, particularly in favelas. After São Paulo, Rio has the largest 'creative GDP' and, of all the businesses within the state, it has the largest number of creative organizations in comparison to the rest of the country (FIRJAN 2016).

The existence of more than 1000 favelas makes Rio de Janeiro a symbol of the informal city, concentrating some of the highest levels of income inequality on the planet. Since their origin, favelas have been the target of stigmatization, persecution and rejection from the 'formal city' that has sought to remove them (Perlman 2010). Although they have specific features and different topographies, morphologies and histories (Varley 2013), favelas or informal settlements can be broadly characterized by an inefficient state provision of basic services and infrastructure; informal relations of work and income generation; and self-built and highly dense housing in environmentally vulnerable locations. They also have lower-than-average economic and educational indicators a plurality of identities and strong social conviviality, with a presence of black, indigenous and migrant populations (Observatório de Favelas 2009). Despite their rich diversity, media and political discourses present them homogeneously as sites of violence, criminality and poverty (Valladares 2005), with a constant war between drug gangs and police.

Creative economy initiatives in peripheral areas have focused on providing entrepreneurial and business training to young people, as in the case of São Paulo or Rio de Janeiro. With the aim of tackling informal practices in the cultural sector, the Cultural Secretariat of the State of Rio de Janeiro created a new incubation agency called Rio Criativo in 2009, working with small creative enterprises through training in business and performance monitoring. In 2014, Rio Criativo launched a programme, Favela Criativa—with support from the private sector and the Inter-American Development Bank—specifically designed to provide arts training, particularly to those aged 15–29, and support cultural entrepreneurship in slums and shanty towns. New funding calls were launched, and business fairs and cultural events with artistic performances were organized. Over a hundred artistic activities took place, involving crafts, cinema, music, theatre and dance performances, fashion and poetry.

Re-branded as the Cultural Territories Programme,[1] the scheme has focused on cultural development for social inclusion in the wider area of the Rio de Janeiro state, and it claims over 2,500 young people have taken part. Formalization training was at the heart of the programme, which offered consultation and advice on how to improve finance, communications and marketing for small enterprises. Temporary cultural infrastructure—such as samba schools, public libraries and sports courts—were set up across various locations in the city's periphery, involving favela-based cultural staff, producers and technicians.

The particularity of Favela Criativa is that, instead of excluding informal cultural practices as most official cultural policies tend to, it made them the target of the programme with the aim of formalizing them. Cultural work in favelas, however,

[1]The programme was expanded so as to include a diversity of locations beyond favelas: rural spaces, urban spaces as a whole and 'quilombos' communities (Secretaria de Cultura RJ 2016). Although the latter originally provided shelter for escaped African slaves, over 3,500 still exist in Brazil (Fundação Cultural Palmares 2016).

has been carried out for a long time without the need of formalization. In the last two decades, the cultural work of Rio-based NGOs such as Nós do Morro, CUFA and AfroReggae has embraced life in the favelas, producing new cultural and urban imaginaries that contest predominant negative social representations about these territories. In so doing, they have reconstituted the relationship between the city and its peripheries. Cultural work in favelas is also central to the reaffirmation of identities in a deeply unequal cartography of cultural infrastructure: 'the creative place of young people from favelas is not a circumstantial act or wandering curiosity, but rather, a way of making culture and making oneself a subject in the city' (Barbosa 2013, p. 23 [author's translation]).

Incorporating individuals and small organizations in the programme also presented difficulties and a lack of administrative resources when it came to account for the financial support received by the programme. Similarly, the Cultura Viva programme, implemented in 2004 under Gilberto Gil's administration as a tool for democratizing cultural policies in Brazil—with its Pontos de Cultura network aimed at decentralizing resources, supporting cultural diversity and putting culture at the service of social inclusion—also operated in peripheral locations, facing similar challenges in post-funding accounting when working with small and informal civil society organizations.

Does the existence of a creative economy programme specifically designed for favelas imply the ultimate subjection of culture and creativity to neoliberal market forces? Most likely yes, in its enactment of a global orthodoxy that equates culture with creativity, creativity with the creative economy and the creative economy with capitalist commercial exchanges. The programme is underpinned by a concern with entrepreneurialism, individual talent and business training, highlighting the need to 'discover' and give value to 'unexploited' creativity. As former Rio State's Culture Secretary Adriana Rattes stated: 'The programme departs from the premise that there is a great and valuable talent pool for arts, culture and creativity in those places [favelas]. And we want to offer the opportunity for them to develop and excel' (original in Portuguese, cited in Gandra 2014). No doubt favela residents are highly entrepreneurial and programmes such as Favela Criativa acknowledged such entrepreneurialism, yet the celebration of ordinary entrepreneurialism is in line with the long history of romanticizing the entrepreneurial flair of slum residents (McFarlane 2012, p. 2798), particularly when it has the potential of expanding and developing new markets.

A relevant example is also the recent music festival 'Rock in Rio 2019'. Here, through the SEBRAE service network, which supports entrepreneurialism, innovation and formalization of small enterprises across Brazil, the festival announced it would be training 10,000 small and medium creative entrepreneurs from the city's favelas. In the words of SEBRAE's president Guilherme Afif Domingos: 'Let's show the value of that [favela] society. It is pointless to make an intervention without promoting social development. The country cannot continue living with the apartheid that exists between the hill [the favela] and the asphalt [the rest of the city], and the way to change this reality is via the creative economy' (original in Portuguese [author's translation], SEBRAE 2018). The hopes pinned on the creative economy for curing existing social and economic exclusions are particularly high in contexts of poverty and deprivation.

The evidence of how sustainable and effective these projects are in the long term remains to be seen. Some of the shortcomings of official initiatives in support of entrepreneurship in favelas like Rocinha, resulted from SEBRAE's prescriptions and entrepreneurship training resources being far from the reality of microentrepreneurs, for they follow different logics. The importance of developing a business plan, carefully deciding the business location or working towards growth and expansion were not among the concerns of entrepreneurs from low-income areas (Pereira et al. 2017).

Creative entrepreneurialism in the favela is also underpinned by political and social goals, seeking to challenge dominant, negative social representations of residents that associate them only with drug-dealing and crime. As Marcus Faustini (the founder of the favela-based NGO Agencia Redes) stated, it is time to start politicizing entrepreneurship instead of adapting creativity to middle-class start-up culture (in Costa and Agustini 2014, p.169). This makes an important distinction between the white, middle-class, university-trained artist from a wealthy part of the city and the young, poor, black cultural worker who lives in a favela and uses cultural production and the arts to reclaim identity and reposition the ideas, visions, perspectives and experiences in the agenda of Brazilian society (Jovchelovitch and Priego-Hernández 2013, p. 52).

Despite existing challenges, Favela Criativa has widened the public visibility of the cultural work of favelas, gathered financial support for popular cultural forms, and provided training to young artists and cultural producers, working with and around informality. 'Formalizing' participants remained a key goal of the programme, which in practice meant participants needed to be on time and work within fixed hours, attend meetings regularly, submit receipts, familiarize themselves with the various tasks involved in cultural production, and learn (formal) patterns of work. While the official support for creativity in informal settlements promotes individual entrepreneurialism by resorting to a commercial rhetoric as well as managerial and subjectivizing discourses of enterprise (Banks 2007), it might also create work opportunities. Formalization, then, can facilitate access to funding and resources, but still has a long way to go to guarantee stable jobs in the highly precarious and informal creative economy markets of Brazil.

7.4 Arte en Barrios, Buenos Aires

In Argentina, the development of creative economy policies has been on the public agenda since 2005, when the government of the City of Buenos Aires created the Metropolitan Design Centre (CMD) in a former fish market that dates back to 1934 in the Barracas neighbourhood. This established a precedent for a number of organizations and initiatives that were then developed at local, national and regional levels. Examples include the Creative Industries General Direction, the

Creative Economy National Secretariat, the specialised market for Argentina's creative industries (MICA) and its regional version MICSUR (the creative industries market of MERCOSUR's countries), and the more recent Creative Cities network as well as the Creative Neighbourhoods programme, inspired by UK initiatives. Being creative—in spite of, or because of, the country's myriad economic crises—became part of the official policy rhetoric and public agenda. This informed a range of fragmented policies aimed at developing entrepreneurialism in artists, cultural producers and citizens more widely, through workshops, training and business accelerator schemes (Del Río 2016; Mérola 2017).

The Buenos Aires municipal government's approach to urban creativity is motivated by urban marketing and socioeconomic development goals. It is epitomized by the launch of the 'creative districts' policy through the creation of a special institutional framework to address existing inequalities in urban infrastructure and resources between the city's wealthier North and its poorer South. Four 'districts' were created between 2005 and 2011 (audio-visual, technology, arts, and design) involving the upgrading of physical infrastructure, the organization of events and festivals, and the provision of trades and technical training free of charge. The districts are expected to function as key tools for economic development as well as urban marketing mechanisms to create, promote and rebrand city areas as geographical clusters of similar economic activity. The government claims that, with the relocation of over a hundred IT companies in a specially designated area in Parque Patricios, the Technology District has led to the creation of 11,000 jobs and a range of physical improvements in urban infrastructure, including the expansion of the underground network and the provision of lighting, new trees and security in the area (Buenos Aires Ciudad 2019).

Although welcome in their improvement of decayed physical infrastructure, these strategies have essentially benefited private companies through a system of public subsidies and tax-exemptions to encourage their relocation. They are driven more by a desire to attract real estate investment and tourism than to pay close attention to the needs of local residents (Bayardo 2013). Similarly, urban regeneration initiatives in the Southern neighbourhoods have mainly focused on embellishment strategies for tourism promotion leading to gentrification. For example, in the case of the Arts District in La Boca, many local artists and activists opposed the municipal government's plans: higher land values and the displacement of the urban poor have been in evidence (Lederman 2015), favouring investors or businesses at the expense of social actors with low purchasing power (Thomasz 2016). As noted by Serafini in her analysis of the cultural and creative industries in Argentina from a post-extractivist perspective (this volume), there is a need to 'move away from understandings of culture that are heavily based on its economic value and to be concerned with social and environmental justice', decolonizing production and decentralizing activities from the city of Buenos Aires. Undoubtedly the creative districts policy faces challenges to better integrate the already established local communities into the planning process and make them the main beneficiaries through education and job opportunities.

A well-known story, this highlights the need to investigate how and whether the promotion of urban creativity can be put at the heart of social justice redistributive agendas. How can it ensure access and participation as well as a distribution of resources, power and opportunities among local communities? Despite critiques of the instrumental nature of cultural policy approaches that use arts projects as agents for social change (Belfiore 2002), access and participation in cultural activities as well as the creative industries have the potential to offer valuable training, develop new skills, create employment opportunities and promote social inclusion (European Commission 2005). In a Southern and Argentinian context this is of particular relevance, where poverty and indigence levels have increased in the last decade (33.6% and 6.1%, respectively in 2018, compared to 25.9% and 4.2% in 2011) (Vera and Salvia 2019), and cultural, artistic and creative activities might help tackle some of the existing economic exclusions, social inequalities and spatial injustices.

In this sense, the launch of a new programme in 2016 for cultural development in the slums and shanty towns of Buenos Aires was a welcome initiative. Named 'Arte en Barrios' (Art in Neighbourhoods), it was implemented by the Under Secretariat of Cultural Management, dependent on the City's Culture Ministry. Argentina has over 4,400 shanty towns where more than 3 million people live (according to the Registro Nacional de Barrios Populares 2018) and the City of Buenos Aires has 42 informal settlements, which are 32 years old on average, and house 82,585 families (TECHO 2016). Most of these settlements have no formal access to the public network of electricity, drinking water or drains. Despite these harsh circumstances, bottom-up, self-managed and collaborative civil society initiatives have created social and cultural engagement initiatives in these territories.

The Arte en Barrios programme—which is a partnership between the Ministry of Culture, the Ministry of Human Development and Habitat, the Secretariat of Social and Urban Integration, and the Housing Institute of the City—has started to develop creative economy and artistic programmes in several informal settlements of the city, particularly in the Southern areas. It has targeted 22 underprivileged neighbourhoods (for example, Villa 21-24, Lugano, Soldati, and Ciudad Oculta) with the aim of delivering weekly trade and artistic workshops, providing training and capacity-building in cultural management, and organizing events and cultural visits to museums and theatres located outside the shanty towns (600 per year approximately, according to official sources). The programme, which mainly attracts women and young people, planned 1,350 activities (including events, projects, visits) during 2019 (personal interview with programme Coordinator, 2019). Both the trades and the artistic workshops have an encompassing scope, covering diverse areas from music production, literature, dance and hairdressing for theatre, to make up, design and screen printing, alongside cultural forms such as cinema, orchestra, percussion and photojournalism. In the words of the city's Culture Minister, Enrique Avogadro: 'Arte en Barrios starts from the premise that, as in all big cities, there is also inequity here. To provide equal opportunities the programme offers training and workshops for 150,000 people living in slums and shanty towns in the city' (Infobae 2019).

Even though the main objective of Arte en Barrios is 'to guarantee universal access to culture, bringing the offer closer to all neighbourhoods of the City of Buenos Aires' (Clarín 2017), this underplays its entrepreneurial component as a mechanism to create and strengthen the cultural labour force of the slums. The work carried out under 'Cultural Management' targets individuals and social organizations who have a particular idea and want to develop it commercially, and the programme provides support and disseminates information about available public funding resources with which the projects could be funded. It is also aimed at the professionalization of artists. In this regard, scholarships are currently being negotiated with a number of educational and cultural institutions (Teatro Colón, Paula Shapiro's Musical Theatre School, Sign Language Music School and Julio Boca's Foundation). Rather than adopting a rigid top-down view that brings the official cultural offer to vulnerable groups from underprivileged settings, the programme puts an emphasis on capacity-building and trade and professional training to improve the integration of the participants into the cultural labour market.

Culture as a tool for social transformation is the underpinning slogan of Arte en Barrios. It was designed to build on the existing territorial work of the Secretariat of Habitat and Inclusion (SECHI) from the City Government. SECHI was created in 2011 as part of the Ministry of Economic Development to coordinate the urbanization policies in shanty towns as well as social, cultural and sports initiatives, previously undertaken by other institutions. Based on ideas of social urbanism and community participation, SECHI collaborated with the Arts in Neighbourhoods programme by providing contacts on the ground and the knowledge and experience of having previously run a community cultural programme ('Cultura Viva') in shanty towns.

The history of public policies for shanty towns in the city of Buenos Aires can be described through three key moments (Brikman 2016, p. 4): eradication policies (1950–1983) from the establishment of shanty towns until the end of the dictatorship; settlement policies (until the early 1990s) supporting dwellers to remain in the territory and to be integrated with the rest of society; and urbanization policies (from 1991 up to the present) seeking to improve living conditions through upgrading initiatives. This is in line with public policy in Latin America and the Caribbean, where informal settlements have been largely accepted and legitimized and improved in various ways (UN-Habitat 2012). While initiatives have included physical interventions in public spaces and community venues, upgrading housing, paving of streets, increased circulation, and basic service provision (mains, electricity, water, etc.), a large proportion has focused on 'urban acupuncture' that seeks to improve the aesthetic features of the neighbourhoods. The programme for the urbanisation of Villa 31 has also included opening a new school and the recent creation of a new cultural venue, Casa de Cultura. More structural and integral responses are needed if policies are to tackle the marginality and precarious access to basic public services and infrastructures in these parts of the city.

The severe economic crisis and state of social emergency that is once again hitting Argentina after its slow recovery from the 2001–2002 crises, which had brought the country into the global spotlight, might lead to redefinitions in urban cultural policy, particularly after the presidential elections in October 2019. Controversially, an additional loan from the International Monetary Fund has resulted in the shrinking of the state, the further plummeting of the national currency, cuts in public spending, labour strikes and street protests—from scientists to transport workers and teachers. This resonates with past unrest resulting from high levels of poverty, rocketing inflation (of 30%), greater social exclusion and urban violence. While this difficult context presents the creative economy field with greater challenges, it also reinforces the need for public policy to support and further develop cultural initiatives for social and economic inclusion in underprivileged areas to ameliorate the negative impact of the crises.

7.5 Conclusion: Supporting Informal Creative Economies in Times of Crises

Visions of the future creative economy are overshadowed by academic pessimism in the North and celebrated by policymakers in the South. In the last two decades, Latin American governments have come to 'discover' the enormous potential that culture and creativity offer for urban and economic development. Many top-down initiatives are following known trajectories, languages and prescriptions from elsewhere; however, there are also some innovative programmes under development, specifically designed for vulnerable groups in informal settlements. Public data, monitoring or evaluation are still very limited in this regard, and it is therefore hard to estimate how many jobs have actually been created as part of the examined programmes.

Engaging with, rather than denying the existence of, the informal economy of culture and creativity offers a good starting point for public policy to develop more inclusive programmes for the sector. Programmes such as Favela Criativa in Rio de Janeiro and Arte en Barrios in Buenos Aires demonstrate that creative economy initiatives can also be designed around social inclusion, socioeconomic development and urban cohesion, even though their politics might derive from an entrepreneurial ethos valuing personal development, self-realization and enterprise discourses (Banks 2007). Still in the early phases of implementation, these programmes have effectively engaged slum and shanty town residents through arts, cultural and crafts training, workshops, and visits to cultural institutions, fairs and events. They have also created links with companies and enterprises operating outside the favelas to connect programme participants with potential employers.

Despite substantial critiques of the creative economy that point to how the creative sector can be exclusionary, non-diverse and elitist—and in so doing reproduce gender, class and racial/ethnic inequalities—this chapter has presented another dimension to the question of exclusion in the creative sector. In Southern contexts of extreme poverty and deprivation, mobilizing the idea of the creative economy can actually generate resources, interest and support that would otherwise not exist, opening up greater possibilities for cultural producers and artists from informal settlements. Their focus on creativity for social inclusion might provide the basis for an alternative model based on skills training, specifically targeting women and young people, support for community cultural activities and events, and facilitating participation in established cultural circuits to develop connections with institutions outside the informal settlements. The creative economy of favelas is based on social innovation, driven by social change and activities that promote citizenship through arts and traditional cultural expressions, and is dependent on networks of small initiatives and enterprises, rather than high technologies, high educational levels and large infrastructures (Schiray et al. 2017).

Although in the future these programmes might not be able to escape a commercial and managerial logic that conflates creativity with profit-making—and some participants might actually welcome such an approach—these initiatives can potentially expand understandings of the creative economy towards promoting social and solidarity economies, based on collaboration between cultural policy and grassroots organizations. More research is needed to understand how these informal economies develop from the bottom-up and to identify ways for public policy to support or strengthen them outside normative and preconceived approaches to the creative economy. For 'what constitutes the economy is a contested terrain, as is the way we see it working to enable or constrain life' (St Martin et al. 2015, p. 1). Certainly, Favela Criativa's or Arte en Barrios' incipient work alone is not enough and has a long way to go to tackle the injustices of cultural work or change the structural conditions of cultural production in shanty towns and slums.

Elsewhere I have shown how investing in the informal creative economy can serve three key functions (Dinardi 2019). From a policy perspective, it can be valued as a market where 'peripheral' cultural forms are given increased market value and are praised as a brand (for example, 'Favela Criativa'). From a social perspective, it can lead to a reconsideration of creativity as social enterprise (McRobbie 2016) based on forms of collective entrepreneurship in the margins and articulated through the significant work carried out by local NGOs on the ground. Finally, it can serve as a reminder of the political nature of creative work as a labour of resistance and radical action, where informality, instead of merely acting as an obstacle or a source of individualization and self-exploitation, inspires creativity, organization and activism. This can be related to the notion of creative justice (Banks 2017) as it sheds light on the prevailing injustices in the cultural and creative industries through a consideration of the values and qualities of cultural objects and, of relevance here, cultural work. Understood as an ethical practice that is socially-embedded in the community, cultural work is part of the moral economy as it involves ethical concerns and a normative judgement about their potentially good contributions to society.

Finally, the creative economy in Latin America relies on a cultural sector that is torn by major economic and political crises alongside disputes over long-standing and unmet needs, such as low or no pay jobs, insufficient public and private sector support, and lack of communication with governmental bodies. It remains to be seen whether the examined policy initiatives in support of the creativity of the less privileged will be able to resist Argentina's severe economic crisis and the far-right forces of Bolsonaro's administration in Brazil, which have aimed to dissolve the Ministry of Culture, restructure cultural funding and censor artists and cultural diversity. Ultimately, the cultural and artistic production of marginalized territories might serve as a platform from which to conceptualize alternative modes of organization and sustainable models of creative economies that support collective entrepreneurship, while also challenging existing forms of exclusion in times of crises.

References

Alsayyad, N. 2004. Urban informality as a 'new' way of life. In *Urban informality: Transnational perspectives from the Middle East, Latin America, and South Asia*, ed. A. Roy and N. Alsayyad, 1–30. Oxford: Lexington Books.

Banks, M. 2007. *The politics of cultural work*. Basingstoke: Palgrave Macmillan.

———. 2017. *Creative justice: Cultural industries, work and inequality*. Maryland: Rowman & Littlefield International.

Banks, M., R. Gill, and S. Taylor. 2013. *Theorizing cultural work: Labour, continuity and change in the cultural and creative industries*. London: Routledge.

Barbosa, J.L. 2013. Favela: Solo cultural da cidade. In *Solos culturais*, ed. J.L. Barbosa and C. Gonçalves Dias. Rio de Janeiro: Observatório de Favelas.

Bayardo, R. 2013. Políticas culturales y economía simbólica de las ciudades: 'Buenos Aires, en todo estás vos.'. *Latin American Research Review* 48: 100–128.

Belfiore, E. 2002. Art as a means of alleviating social exclusion: Does it really work? A critique of instrumental cultural policies and social impact studies in the UK. *International Journal of Cultural Policy* 8 (1): 91–106.

Brikman, D. 2016. ¿Gestión social de Hábitat? La política del PRO en las Villas de CABA, 2011–2015. *Quid 16, IIGG* 6: 1–26.

Brillembourg, A., H. Klumpner, and A. Kalagas. 2015. El Sur global: Future resilient city. *Architectural Design* 85 (4): 100–105.

Brook, O., D. O'Brien, and M. Taylor. 2018. Panic! Social class, taste and inequalities in the creative industries. https://static.a-n.co.uk/wp-content/uploads/2018/05/Panic-Social-class-taste-and-inequalities-in-the-creative-industries.pdf. Accessed 13 May 2019.

Buenos Aires Ciudad. 2019. Conocé el Distrito Tecnológico de la Ciudad.' Buenos Aires BA website. https://www.buenosaires.gob.ar/noticias/conoce-el-distrito-tecnologico-de-la-ciudad. Accessed 13 May 2019.

Buitrago, F., and I. Duque. 2013. *La economía naranja: Una oportunidad infinita*. New York: Banco Interamericano de Desarrollo.

Clarín. 2017. De barrios vulnerables a la Usina del Arte, para mostrar y festejar. *Clarín*, December 17. https://www.clarin.com/ciudades/barrios-vulnerables-usina-arte-mostrar-festejar_0_r1GWjKEzG.html. Accessed 13 May 2019.

Conor, B., R. Gill, and S. Taylor. 2015. Gender and creative labour. *The Sociological Review* 63 (1_suppl): 1–22.

Costa, E., and G. Agustini, eds. 2014. *De baixo pra cima*. Rio de Janeiro: Aeroplano editora.

Curtin, M., and K. Sanson. 2016. *Precarious creativity: Global media, local labor*. Berkeley: University of California Press.

De Souza e silva, J. 2014. As periferias roubam a cena cultural carioca. In *De baixo para cima*, ed. E. Costa and G. Agustini, 55–78. Rio de Janeiro: Aeroplano.

Del Río, J. 2016. Emprendedores y Gobierno, fin de la grieta. *La Nación*, March 13. https://www.lanacion.com.ar/economia/emprendedores-y-gobierno-fin-de-la-grieta-nid1879108. Accessed 13 May 2019.

Dinardi, C. 2019. Creativity, informality and cultural work in Rio de Janeiro's favelas. *International Journal of Cultural Studies* 22 (2): 248–263.
European Commission. 2005. The role of culture in preventing and reducing poverty and social exclusion. *Policy Studies Findings* 2: 1–8.
FIRJAN. 2008. *A cadeia da indústria criativa no Brasil*. Rio De Janeiro: Firjan.
———. 2016. Indústria criativa. Análise especial: Rio de Janeiro. FIRJAN. http://www.firjan.com.br/economiacriativa/download/analise_especial_rj_-_2016.pdf. Accessed 13 May 2019.
———. 2019. Mapeamento da indústria criativa no Brasil. FIRJAN. https://www.firjan.com.br/EconomiaCriativa/downloads/MapeamentoIndustriaCriativa.pdf. Accessed 13 May 2019.
Fundação Cultural Palmares. 2016. Quilombos ainda existem no Brasil. *Palmares Fundação Cultural*. http://www.palmares.gov.br/?p=3041. Accessed 13 May 2019.
Gandra, A. 2014. Favela Criativa desenvolve talentos em comunidades do Rio. *Exame*, June 26. https://exame.abril.com.br/brasil/favela-criativa-desenvolve-talentos-em-comunidades-do-rio/. Accessed 13 May 2019.
Gill, R. 2014. Unspeakable inequalities: Post feminism, entrepreneurial subjectivity, and the repudiation of sexism among cultural workers. *Social Politics* 21 (4): 509–528.
Gill, R., and A. Pratt. 2008. In the social factory? Immaterial labour, precariousness and cultural work. *Theory, Culture and Society* 25 (7–8): 1–30.
Hesmondhalgh, D., and S. Baker. 2011. *Creative labour: Media work in three cultural industries*. New York: Routledge.
Infobae. 2019. ¿Cuáles son las estrategias culturales de México, Londres, Washington D.C. y Lisboa? *Infobae*, May 3. https://www.infobae.com/america/cultura-america/2019/04/03/cuales-son-las-estrategias-culturales-de-mexico-londres-washington-d-c-y-lisboa/. Accessed 13 May 2019.
International Labour Organization. 2018. Women and men in the informal economy: A statistical picture. Third edition. ILO.
Jauregui, J.M. 2011. Articulating the broken city and society. *Architectural Design* 81 (3): 58–63.
Jovchelovitch, S., and J. Priego-Hernández. 2013. *Underground sociabilities: Identity, culture and resistance in Rio de Janeiro's favelas*. Brasilia: UNESCO.
Landry, C. 2000. *The creative city: A toolkit for Urban Innovators*. London: Earthscan.
Lederman, J. 2015. Urban fads and consensual fictions: Creative, sustainable, and competitive city policies in Buenos Aires. *City Community* 14 (1): 47–67.
Littler, J. 2017. *Against Meritocracy*. London: Routledge.
Mbaye, J., and C. Dinardi. 2019. Ins and outs of the cultural polis: Informality, culture and governance in the global South. *Urban Studies* 56 (3): 578–593.
McFarlane, C. 2012. The entrepreneurial slum: Civil society, mobility and the co-production of urban development. *Urban Studies* 49 (13): 2795–2816.
McRobbie, A. 2016. *Be creative: Making a living in the new culture industries*. Cambridge: Polity Press.
Mehrotra, R. 2010. Foreword. In *Rethinking the informal city: Critical perspectives from Latin America*, ed. F. Hernández, P. Kellett, and L.K. Allen, xi–xiv. New York: Berghan Books.
Merkel, J. 2018. 'Freelance isn't free.' Co-working as a critical urban practice to cope with informality in creative labour markets. *Urban Studies* 56 (3): 526–547.
Mérola, M. 2017. ¿Nueva política pública o moda pasajera? El caso de la política de desarrollo emprendedor de la Ciudad Autónoma de Buenos Aires entre 2008 y 2015. MA Thesis, University of Georgetown. https://repository.library.georgetown.edu/bitstream/handle/10822/1044654/Merola_georgetown_0076M_13757.pdf?sequence=1&isAllowed=y. Accessed 13 May 2019.
O'Brien, D., and K. Oakley. 2015. *Cultural value and inequality: A critical literature review*. Swindon: AHRC.
O'Brien, D., D. Laurison, A. Miles, and S. Friedman. 2016. Are the creative industries meritocratic? An analysis of the 2014 British Labour Force Survey. *Cultural Trends* 25 (2): 116–131.
Observatório de Favelas. 2009. *O que é a favela, afinal?* Rio de Janeiro: Observatório de Favelas.

Palma, J., and M. Garcia Lozano. 2017. Building more affordable and disaster-resilient housing in Latin America and the Caribbean: A few policy ideas. Sustainable Cities: *The World Bank blog*, May 4. https://blogs.worldbank.org/sustainablecities/building-more-affordable-and-disaster-resilient-housing-latin-america-and-caribbean-few-policy-ideas. Accessed 13 May 2019.

Pereira, I., R. Bartholo, É.R. Silva, and D. Proença. 2017. Entrepreneurship in the favela of Rocinha, Rio de Janeiro: A critical approach. *Latin American Research Review* 52 (1): 79–93.

Perlman, J. 2010. *Favela: Four decades of living on the edge in Rio de Janeiro*. Oxford: Oxford University Press.

Registro Nacional de Barrios Populares. 2018. Relevamiento Nacional De Barrios Populares, available online: https://www.argentina.gob.ar/habitat/renabap. Accessed 13 May 2019.

Saha, A. 2017. *Race and the cultural industries*. Cambridge: Polity Press.

Schiray, D.M., C.C. Carvalho, and R. Afonso. 2017. Creative economy as a social technology approach: A case study in favela da Mangueira, Rio de Janeiro, Brazil. *Academia: Revista Latinoamericana de Administración* 30 (4): 508–528.

Schlesinger, P. 2017. The creative economy: Invention of a global orthodoxy, Innovation. *European Journal of Social Science Research* 30 (1): 73–90.

SEBRAE. 2018. Sebrae levará empreendedores das favelas para o Rock in Rio 2019. *Sebrae News portal*, April 25. http://agenciasebrae.com.br/sites/asn/uf/NA/sebrae-levara-empreendedores-das-favelas-para-o-rock-in-rio-2019,a2465c6932df2610VgnVCM1000004c00210aRCRD. Accessed 13 May 2019.

Secretaria de Cultura, R.J. 2016. Inscrições para edital Cidade Criativa Incubação. http://cultura.rj.gov.br/. Accessed 13 May 2019.

St. Martin, K., G. Roelvink, K. Gibson, and J. Graham. 2015. Introduction: An economic politics for our times. In *Making other worlds possible: Performing diverse economies*, ed. G. Roelvink, K.St. Martin, and J.K. Gibson-Graham, 1–25. Minneapolis: University of Minnesota Press.

TECHO. 2016. Relevamiento de asentamientos informales. TECHO Argentina. https://www.techo.org/argentina/wp-content/uploads/sites/3/2019/04/Informe-Relevamiento-de-Asentamientos-Informales-2016-TECHO-Argentina.pdf. Accessed 13 May 2019.

Thomasz, G. 2016. Los nuevos distritos creativos de la Ciudad de Buenos Aires: la conversión del barrio de La Boca en el 'Distrito de las Artes'. *EURE (Santiago)* 42 (126): 145–167.

UNCTAD. 2008. Creative economy report: The challenge of assessing the creative economy. Towards informed policy-making. United Nations. https://unctad.org/en/Docs/ditctab20103_en.pdf. Accessed 13 May 2019.

———. 2010. Creative economy report: Creative economy, a feasible development option. United Nations. https://unctad.org/en/Docs/ditctab20103_en.pdf. Accessed 13 May 2019.

———. 2013. Creative Economy report 2013. Special edition: Widening local development paths. United Nations. http://www.unesco.org/culture/pdf/creative-economy-report-2013-en.pdf. Accessed 13 May 2019.

———. 2018. Creative economy outlook: Trends in international trade in creative industries. United Nations. https://unctad.org/en/pages/publications/Creative-Economy-Report-(Series).aspx. Accessed 13 May 2019.

UN-Habitat. 2012. State of Latin American and Caribbean cities 2012: Towards a new urban transition. UN Habitat. https://unhabitat.org/books/state-of-latin-american-and-caribbean-cities-2/. Accessed 13 May 2019.

Valladares, L.D.P. 2005. *A invenção da favela: Do mito de origem a favela.com*. Rio de Janeiro: Editora FGV.

Varley, A. 2013. Postcolonialising informality? *Environment and Planning D: Society and Space* 31 (1): 4–22.

Vera, J., and A. Salvia. 2019. Pobreza y desigualdad monetaria en los hogares urbanos de la Argentina a partir de la encuesta de la deuda social argentina (2010–2018). Working paper, UCA. Buenos Aires: UCA.

Chapter 8
Towards a New Paradigm of the Creative City or the Same Devil in Disguise? Culture-led Urban (Re)development and Sustainability

Mariangela Lavanga and Martina Drosner

8.1 Introduction

With an expected 68% of the world's population living in a city by 2050 (United Nations Department of Economic and Social Affairs 2018), climate change and rapid global urbanization are among the great challenges of our time, exerting economic, social and environmental pressure on cities worldwide. The need for new models of urban planning and development in order to create open, safe, resilient and sustainable cities has been acknowledged by the world community as part of the 2030 Agenda for Sustainable Development (ASD) and its 17 Sustainable Development Goals (SDGs), which were adopted by the UN General Assembly in 2015 (Throsby 2017; United Nations 2019). As we witness the emergence of a paradigm shift within urban planning, the relationship between culture, the economy and cities needs to be rethought, not only with regard to the 2030 ASD, but similarly in relation to building creative metropolises that are also sustainable and circular.

The Brundtland report defines sustainable development as 'the development that meets the needs of the present without compromising the ability of future generations to meet their own needs' (WCED 1987, p. 41). The report used the concept of sustainable development to particularly link economic and environmental systems. It also stressed the long-term properties of the concept of sustainability, and introduced 'the ethical principle of achieving equity between the present and the future generations' (Diesendorf 2000, p. 3). A 'three pillars model' of sustainability then emerged, which acknowledged environmental, financial and social qualities (United

The original version of this chapter was revised. The correction to this chapter is available at https://doi.org/10.1007/978-3-030-49384-4_12

M. Lavanga (✉) · M. Drosner
Erasmus University Rotterdam, Rotterdam, The Netherlands
e-mail: Lavanga@eshcc.eur.nl

K. Oakley, M. Banks (eds.), *Cultural Industries and the Environmental Crisis*,
https://doi.org/10.1007/978-3-030-49384-4_8

Nations World Summit 2005).[1] References to culture and its role were, however, underrepresented or missing in both academic and policy discourses. In fact, only a few scholars have recognised the part it plays. Cultural economist Throsby (1995, 2008, 2017) has long argued that it is not possible to speak about sustainability without also talking about culture. Hawkes (2001), meanwhile, discussed four pillars of sustainable development, acknowledging that environmental, financial and social sustainability require a foundation in culture if they are to be understood and implemented at an individual level. According to Hawkes, cultural vitality is as important for a sustainable society as 'social equity, environmental responsibility and economic viability' (Hawkes 2001, p. vii). The 2008 financial crash and recent environmental crises have intensified the debate around sustainable development and a circular economy, but the role of culture is still undervalued. Indeed, the 2030 ASD, and SDG 11 (Sustainable Cities and Communities) in particular, does not mention culture's part in sustainable urban development at all (Throsby 2017).

In this chapter, we focus on the relationship between culture and sustainable development. To do this, we look at the case of De Ceuvel, a planned work space for creative and social enterprises, which is located in a polluted area in a former shipping wharf in Amsterdam North (the Netherlands). De Ceuvel's location is heavily polluted and has been excluded from the municipality's masterplan ("ground exploitation programme" under the new Dutch Spatial Law) for 10 years. A tendering process was, however, developed by the Municipality of Amsterdam and other stakeholders as a way to offer the land to interested parties. The objective was to combine culture with environmental sustainability, in particular, and provide new affordable work settings for cultural and creative entrepreneurs. After winning the tender in 2012, the Amsterdam-based architecture bureau Space&Matter and the architect Marjolein Smeele (Smeelearchitecture agency) set up an expert group to plan the regeneration of the area. Working with cultural and creative entrepreneurs that moved there, they transformed De Ceuvel into a clean-tech playground with the aim of making 'sustainability tangible, accessible and fun' (De Ceuvel 2019). As an experimental setting, De Ceuvel may be particularly relevant concerning the ways in which culture is interrelated with sustainable development. Our goal in this chapter is to uncover what inspires cultural and creative entrepreneurs to commit to sustainable development, and what their intentions, ambitions and visions are. Given the focus of the public tender on culture and the environmental sustainability pillar, we anticipate that this interrelationship may not be enough to promote a genuine, long-term sustainability-oriented project.

Data on the De Ceuvel case was collected in 2015 using qualitative semi-structured interviews and participant observations on site. This was complemented with a content analysis of De Ceuvel's website, the tender of the Municipality of Amsterdam, De Ceuvel's application documents, and policy documents on urban development plans for Amsterdam North. In order to depict recent developments

[1] Economic sustainability refers to the efficient and responsible use of resources by businesses, so that they can operate and be profitable. Environmental sustainability means the use of natural resources in a way that considers scarcity and the needs of future generations. Socially sustainable communities are resilient and capable of adapting to change.

and changes, we conducted a further analysis of the situation in 2019, which was based on structured, written interviews with both the tenants who had been interviewed in 2015 and new arrivals.[2] Table 8.1 contains further information on the interviewees, who were all Caucasian and well educated (they all had master's degrees, save for one who had a bachelor's degree).

This chapter addresses the role of culture in sustainable development. It examines the origins of De Ceuvel and focuses on cultural and creative entrepreneurs and their decisions to commit to sustainable development. It also considers the future of De Ceuvel and assesses to what extent it is a culturally sustainable development experiment.

Table 8.1 List of interviewees

Interviewee #	Gender	Age range	Co-founder/ tenant	School degree	Profession	Interview year
1	M	25–35	Assistant to co-Founder	MSc	Consultant at consultancy in circular economy	2015, 2019
2	M	35–45	Tenant	MA	Craftsman	2015
3	M	25–35	Founder	MA	Architect	2015
4	M	35–45	Tenant	n.a.	Craftsman	2015
5	M	35–45	Tenant	BA	Craftsman	2015
6	F	35–45	Tenant	MSc	Futurologist	2015 (no longer a tenant in 2019)
7	F	35–45	Tenant	MA	Spatial designer	2015, 2019
8	M	35–45	Tenant	MSc	Architect	2015, 2019
9	M	35–45	Tenant	MSc	Project manager at creative content agency	2015 (no longer a tenant in 2019)
10	M	35–45	Founder	MSc	Architect	2015
11	F	35–45	Tenant	MSc	Architect and interior designer	2019
12	M	35–45	Municipality of Amsterdam	MSc	Project manager	2015
13	F	45–55	Municipality of Amsterdam	MA	Project manager	2015

[2] The first sample consisted of 12 interviewees (eight tenants, two founders and two project managers from the City of Amsterdam) and the second contained four (all tenants). Three of the tenants interviewed in 2015 had left De Ceuvel between the two data collection rounds. In addition, a content analysis of De Ceuvel's website and its annual reports from 2014 to 2017 complemented the investigation. The data collected in 2015 was part of the thesis "Sustainability and Culture" written by Martina Drosner for her MA in Cultural Economics and Entrepreneurship (Erasmus University Rotterdam). A previous version of this chapter was presented at the AAG (Association of American Geographers) annual conference in San Francisco in 2016, as part of the session "Negotiating the 'creative city': A new cultural governance for the 'creative city'?"

8.2 Culturally Sustainable Development

Culture-led urban development policymaking has come a long way. Indeed, in the context of a declining post-industrialized economy, bottom-up regeneration has taken place in derelict industrial brownfields like Tacheles in Berlin and Friche la Belle de Mai in Marseille (Andres and Grésillon 2013). In many cities worldwide, especially in Western countries, artists and creative entrepreneurs were attracted to former industrial buildings as part of their search for affordable, low-maintenance, urban spaces that could be adapted to their individual requirements (Zukin 1989). Organic and decentralized, these cultural activities grew into spaces that combined cultural production and cultural consumption facilities with little or no involvement from local or national governments. With the rise of cultural and creative industries in the 2000s, policymakers aimed to capitalize on cultural production as a way to stimulate economic growth (Braun and Lavanga 2007; Evans 2009; Foord 2009; Lavanga 2004, 2013; van der Borg et al. 2005). Top-down creations of cultural districts, quarters or clusters were expected to boost the economy and revitalize urban areas (Lavanga 2020), with copy and paste solutions circulating worldwide (Pratt 2008).

Scholars have strongly criticized cultural policies that drifted towards more market-driven redevelopments, in particular for the lack of attention paid to place-specific characteristics and the increase in socio-economic and spatial inequalities (for example, Banks 2017; Banks and O'Connor 2009; Hesmondhalgh et al. 2014; Lavanga 2006, 2009, 2013; Oakley 2004; Oakley and O'Connor 2015; Rozentale and Lavanga 2014; van der Borg et al. 2005). Culture has mostly been used for instrumental and financial ends for the past few decades, and cultural industries have long been praised for their contributions to innovation and economic growth. Their influence on non-material wellbeing or 'the good life and good society' (Klamer 2002, p. 470) has, however, often been neglected.

The concept of culturally sustainable development may be useful for reconnecting and reconciling culture with the economic, societal and environmental pillars. According to Throsby, sustainable development based on culture 'encompasses both the idea of cultural development in its own right, according art and culture an independent and valued role in their own terms within society, and culture as a set of attitudes and practices that can be instrumental in supporting, constraining and/or contributing to economic and social development in the widest sense' (Throsby 1995, p. 202). Two main definitions of culture can be derived: (a) functional, which refers to the cultural and creative industries and (b) anthropological, which references the intangible values, beliefs, rituals and traditions of a society. Thorsby suggested four criteria for defining and assessing culturally-sustainable development. First, development cannot be measured by economic growth alone, but should also address the advancement of material and non-material wellbeing. Along with economic indicators, environmental, social and cultural versions must be added too. Second, a long-term view and responsibility towards future generations are essential. Indeed, sustainable development should strive for intergenerational equity—'fairness in the distribution of resources and opportunities between generations, in particular between present and future generations' (Ibid., p. 203)—and the maintenance of environmental and cultural capital (in its material and immaterial forms). Third, individuals should have the right to fair treatment in the economic, social and cultural spheres. Intragenerational equity means a fair distribution of economic and social opportunities and resources, along with fair access to cultural resources for disadvantaged groups of the present generation. Finally, we need to recognize the interdependence between economic, environmental, social and cultural processes as fundamental for assessing sustainable development. In conclusion, a system-wide approach to policymaking is therefore required.

Moving from a macro- to a meso-level, Kirchberg and Kagan (2013, p. 141) provide insight into what a sustainable, creative city may represent. They argue that this is a place where people can 'practice ecologically resilient, socially equitable and inter-culturally vibrant modes of life.' Several elements are recognised as essential: the practice of collaborative art; an openness to experimenting with alternative locally and globally sustainable ways of life; artisanship to complement communal practice; and recognizing the arts as a catalyst for integration, which in turn stimulates the exchange of different cultures within a community (Kagan and Hahn 2011). A creative, sustainable place may share some of the features of utopian experiments and inspire change in society at large. Such experiments, whether it be Auroville in India or Christiania in Copenhagen, have suggested that self-sustaining communities may offer creative solutions to environmental and societal issues (for example, de Geus 1999; Sargent 2005; Sargisson 2000; Vanolo 2013). Utopias may serve to motivate people to make socially and environmentally sustainable choices and reassess their way of life: for example, by enabling a process of holistic thinking about what is a 'good society (Levitas 2000); by 'doing the right thing' and uncovering their true values (Klamer 2017); by inspiring alternative visions of the 'good life' (Oakley and Ward 2018); or by fostering an 'ecological citizenship' (Duxbury et al. 2017).

While the conceptualisation of culturally sustainable development is advancing, operationalising it is still difficult. Moving towards a more micro-level, we aim to explore why and how cultural and creative entrepreneurs commit to the concept. In doing so, our goal is to examine to what extent De Ceuvel is an experiment with this process.

8.3 The Origins of De Ceuvel

De Ceuvel is an eco-hub for creative and social enterprises and is located in a former industrial area of Amsterdam North. It opened to the public in 2014 and describes itself as 'a city playground for innovation, experimentation and creativity.' It aims 'to make sustainability tangible, accessible and fun' (De Ceuvel 2019). De Ceuvel differs from other Dutch urban redevelopment projects in several ways: (a) its location in a heavily polluted former ship wharf; (b) its innovative approach to turning an urban brownfield into a self-sustaining 'regenerative urban oasis' by way of integrated clean technologies (Ibid.); (c) the fact that it was stimulated top-down via a tender, but developed organically with the aim of attracting creative and social entrepreneurs interested in sustainable development; and (d) its combination of culture and sustainable development at various levels, which is motivated by the belief that 'the transition to a circular economy and society is not only a technical transition, it's also a cultural transition' (Ibid.).

Considering the brief history of De Ceuvel sheds light on its outcomes as it developed. Its narrative begins with its top-down establishment via the Municipality of Amsterdam. However, for two reasons, its evolution differs from the standard culture-led regeneration strategies identified by Lavanga (2004, 2009) and Andres and Grésillon (2013). First, the urban redevelopment plans for Amsterdam North were put on hold due to the financial crisis. As a result, the land remained undeveloped, as was also the case with the former ship wharf Ceuvel Volharding. Second, this particular piece of land in the Buiksloterham area was further excluded from the local government's ground exploitation programme, because the soil is heavily polluted due to the former industrial production of ships. Consequently, no revenue was expected for 10 years. As stated in the report 'Circular Buiksloterham', 'its polluted lands and open spaces can become the centre of the implementation of new clean technologies and a hub for the closure of urban material cycles […] As such, Buiksloterham can serve as a blueprint and live experiment for how such formerly peripheral areas worldwide can be transformed into a motor for change and regeneration in cities' (Metabolic, Studioninedots and DELVA Landscape Architects 2014, p. 4–5).[3] In order to utilize the land, a competition for its temporary use was launched in collaboration with three government departments: Stadsdeel Amsterdam Noord (Amsterdam North City District, the owner of the land); Projectbureau Noordwaarts (responsible for the redevelopment of the post-industrial area of Buiksloterham); and Bureau Broedplaatsen, whose function was to create an 'affordable workspace for artists in the city'[4] (Interviewee 12, policymaker, 2015). In a city fascinated by Florida's theory of the creative class, and which is experiencing gentrification and over-tourism, Bureau Broedplaaatsen has been in charge of keeping alive the creative, often alternative and anarchist, vibe and infrastructure that makes Amsterdam renowned worldwide (van der Borg et al. 2005).

The three city-government departments created the tender to make the land available for a creative breeding space and offer cheap places to work to cultural and creative entrepreneurs. Excluded for 10 years from the municipality's masterplan

[3] According to Metabolic, Studioninedots and DELVA Landscape Architects (2014), the polluted plots in Buiksloterham covered over 15 hectares, circa 15% of the total area.

[4] Bureau Broedplaaatsen is a department of the Municipality of Amsterdam. It is in charge of developing creative 'broedplaatsen, creative breeding grounds or creative locations which host affordable work spaces for artists and creative entrepreneurs in Amsterdam. A diverse range of stakeholders works with Bureau Broedplaatsen to develop the creative broedplaatsen, ranging from artist groups to housing corporations and creative management organisations. It was formed as a follow-up to the BroedplaatsAmsterdam project.

("ground exploitation programme" under the new Spatial Law), the polluted land was offered to interested parties with an objective of combining both culture and environmental sustainability in the creation of a new cultural breeding space. Potential tenants had to fulfil a number of criteria, as set out in Table 8.2.

The Amsterdam-based architecture bureau Space&Matter and the architect Marjolein Smeele (Smeelearchitecture agency) won the tender in 2012. They brought together a team of experts ranging from architects involved in design, project development and management, to organizations specialising in concept development, research, the implementation of clean technology and financial services.[5] With the land secured with a 10-year lease, the team worked together to transform the former de Ceuvel Volharding shipyard into a place to experiment with a circular economy, clean technology, urban development and culture.

De Ceuvel's spatial structure consists of refurbished houseboats turned into workspaces and studios. These are located on the polluted soil and are equipped with circular and clean technologies. A wooden jetty connects the houseboats to prevent contact with the soil. Regenerative landscape architecture was also used for the so-called 'forbidden garden', where 'phyto-remediating plants' extract pollutants from the soil, thereby cleansing it (De Ceuvel 2019). This spatial arrangement is complemented by De Ceuvel's soft structure, which consists of its tenants (the creative and social entrepreneurs) and a hospitality facility. As one founder explained, they had the vision to '… show in a way that we can combine sustainability and culture in De Ceuvel by inviting people to cultural events' and 'by also showing them new ways and technologies for sustainability' (Interviewee 10, founder, 2015).

Table 8.2 The tender selection criteria (municipality of Amsterdam)

	Criteria considered	Weight (%)
1.	The need to develop a concept that is extraordinary and/or of high quality with regard to innovative ideas about how environmental sustainability is incorporated	40
2.	The degree to which the concept is technically and financially feasible; it must also adhere to the requirements imposed by the departments responsible for both the area's urban redevelopment and supporting creative "broedplaatsen" in Amsterdam	30
3.	The extent to which the concept proves to be a qualitative addition to the physical environment, including concerning quality of life and social security	15
4.	The degree to which the applicants are experienced in the operation of creative "broedplaatsen", real estate and/or hospitality	15

Source: Authors' explanations based on the document 'Prijsvraag Broedplaats Ceuvel Volharding'

[5] The complete list of organizations is available on De Ceuvel's website and includes: Space&Matter (urban planning, architecture, project development); Smeelearchitecture (project development, community building); the architect Jeroen Apers (project development, finances); Marcel van Wees (project and general management); Metabolic (concept development, research and implementation of clean technologies and sustainability plans); DELVA Landscape Architects & Bureau Fonkel (design, implementation and maintenance of the Purifying Park); Studio Valkenier (designed Café de Ceuvel); and Waterloft (financial advice).

This vision, which was paired with a creative problem-solving approach, was what convinced the jury, as explained by one of its members:

> What's really important at De Ceuvel is not that they actually leave the ground cleaner. That's nice, but I think the BIG thing is that you can show people what's possible. Even on a tight budget, and even if you aren't investing a lot of money, you can still do things by having ideas, thinking about it in a smart way. I think that's the big thing that De Ceuvel shows. Not so much the result, but instead showing people the possibilities. (Interviewee 10, founder, 2015)

De Ceuvel's location has been redeveloped with a considerable voluntary effort from its tenants via the creation of the Association de Ceuvel (ADC). This is currently comprised of its board (chair, treasurer, secretary and two tenants), four commercial boats (Crossboat, Metabolic Lab, Café de Ceuvel, Hotel Asile Flottant)[6] and 14 non-commercial boats that house a mini theatre and studio, ateliers, and workspaces for artists and creative entrepreneurs. Boats can either be rented individually or shared.[7]

8.4 Now and the Future of De Ceuvel, or the Temporality Paradox

De Ceuvel accommodates 26 companies, entrepreneurs and artists. The rest of the space is occupied by a floating hotel, and three boats are available for rent for (public) events, meetings and workshops. The plot also has a café, which is environmentally sustainable in three ways: it was built using upcycled materials; it works with local organic food suppliers; and it grows its own herbs in its rooftop greenhouse. Most of the tenants work in the fields of design, social innovation and architecture.[8] The rest are involved in visual art, communication, and the performing arts, while a minority are active in crafts, journalism and event management. Our sample in Table 8.1 is representative of the current population. Architecture and spatial design are the most represented sectors. The majority of our 11 creative entrepreneurs, including the founders, were male and aged between 35 and 45. A quick scan of the current tenants on De Ceuvel's website provides a snapshot of these middle-class, highly-educated, Caucasian entrepreneurs.

Our interviewees mentioned three elements they regarded as important when deciding whether to move to De Ceuvel: first, the community spirit and networking

---

[6] Crossboat and Metabolic Lab host workshops and events. Hotel Asile Flottant is comprised of six boats retrofitted as hotel rooms, and is docked on the canal at De Ceuvel. It is an independent organisation and a member of the Association de Ceuvel (De Ceuvel 2019).

[7] For an overview of the current tenants and boats, please visit https://deceuvel.nl/en/boats/. See Table 8.1 for a list of the tenants and founders interviewed.

[8] The category 'design' includes different aspects, such as landscaping and interior and graphic design. 'Social innovation' refers to activities concerned with designing and developing solutions to meet social needs.

opportunities; second, the amenities (the café, the aesthetics, the nature and the cultural activities); and third, major technological innovation in terms of environmental sustainability. Being part of a group 'that works on innovative sustainable projects in Amsterdam' (Interviewee 8, tenant, 2019) seemed to be closely connected to the third factor. The same tenant pointed out, 'it feels like we're at the forefront of [environmentally] sustainable innovation, which is the place I'd like to be.' Another tenant stressed the importance of the like-minded community feeling:

> For me [it was] the idea of having your own studio with (...) a group that's got the same mind-set about taking this experience of making a sustainable, closed-looped city area, which is quite experimental. (Interviewee 7, tenant, 2015)

More than 5 years after opening to the public in 2014, De Ceuvel is seen as 'one of Europe's most unique urban experiments' (Morton 2018). It has received much international media exposure and has won awards and nominations worldwide. Its development is an 'ongoing process' on different levels (Interviewee 7, tenant, 2019) but is always aligned with the vision of environmental sustainability. Among the various highlights are the newly-built aquaponics greenhouse, which uses 'a closed-loop aquaponics system combining fish and vegetable production' to grow herbs and vegetables for Café de Ceuvel (De Ceuvel 2019); the introduction of Juliette, a blockchain-based solar trading system, which fuels local production and the exchange of renewable energy locally;[9] the opening of the floating Hotel Asile Flottant;[10] and the twofold extension of Café De Ceuvel, due to its increased popularity with the public (Interviewee 11, tenant, 2019).

Nevertheless, not all of the ADC's environmental sustainability projects have come to fruition. The Bio-gas Boat, for instance, was planned as a way to 'convert organic waste into biogas' that could be used in the café (De Ceuvel 2019). However, the project ended in 2019 after technical difficulties caused the ship to sink during a storm (Interviewee 7, tenant, 2019). It did, however, inspire those involved in a sanitation project in the wider area of Buiksloterham, which was carried out by the local Amsterdam water company.

One of the challenges facing De Ceuvel concerns the future of the ADC. As one tenant pointed out, 'After the first 5 years, we're looking to the future to see what De Ceuvel can mean after the end of the 10-year term' (Interviewee 7, tenant, 2019). The experiment has proceeded on a 'trial and error' basis, which makes it clear that such a development will be a long-term process. This raises the question of how easily, and to what extent, De Ceuvel can be seen as a model for other projects. Accordingly, the tenants still have a vision to 'recreate it [De Ceuvel] somewhere else' (Interviewee 8, tenant, 2015); indeed, this is behind the current conversations and negotiations the ADC is having with the Municipality of Amsterdam. De Ceuvel

[9] 'Mining existing crypto-currencies such as Bitcoin consumes a lot of energy, while the Jouliette is generated by the user producing excess solar energy … Jouliette encourages solar panel owners to exchange energy locally, instead of selling surplus power to the grid' (De Ceuvel 2019).

[10] Asile Flottant is a member of the ADC, but works as an independent business and pays a small fee for maintenance of the site.

could be recreated at a different location, as the houseboats can be moved, thus exemplifying to the public and the local government that creative, sustainable solutions can be designed in a way that outlasts temporal and spatial restrictions. Houseboats have been a common sight on many canals in Amsterdam since the 1970s, and their prices are competitive with those of traditional homes. De Ceuvel works as a community of houseboats that can be moved to other former docks along the River IJ, countering the lack of affordable work and housing spaces in the city centre.[11] In this respect, De Ceuvel may encapsulate a temporality paradox: it was born as a temporary project, but can be a permanent one, too. Its temporality, combined with its reliability in terms of its mobile infrastructure, may thus enable De Ceuvel to colonize other fringe areas in Amsterdam that need clean-tech innovations. However, the challenge of the De Ceuvel experiment is not to simply replicate and recreate it in a new location, but to adapt to a new context and related economic, environmental, social and cultural challenges.

8.5 Culture and Environmental Sustainability at De Ceuvel

The relationship between culture and environmental sustainability takes at least three different forms at De Ceuvel. First, it was policy-induced and planned via its top-down implementation as a creative 'broedplaatsen' and a 'playground for sustainable technologies' (De Ceuvel 2019). Following the redevelopment of the area led by experts, the first tenants arrived and the ADC opened its premises up to the broader public. As one tenant explained, the effects of the policy are still significant:

> That's not because they [creative entrepreneurs and artists] would be more inclined to go for [environmental] sustainability. It's just because the Bureau Broedplaatsen has a policy where you have to be a cultural entrepreneur to be here. ... Of course, from a general, demographic perspective, this [group of entrepreneurs] would be an educated group, so we make decisions that are more conscious. But it's not that we're here because of it [environmental sustainability]. (Interviewee 2, tenant, 2015)

Second, the connection between culture and environmental sustainability may not have been inherent in the tenants' work initially but could have grown over time in parallel with the increasing global, national and local interest in sustainable development and a circular economy. One tenant summarised that:

> [a] lot of the cultural creatives have [environmental] sustainability as a theme in their work, and a lot of the sustainable ecopreneurs need creative thoughts to bring their ideas to life. I think they go hand in hand. (Interviewee 8, tenant, 2019)

[11] Space&Matter is also involved in the development of Schoonschip in Amsterdam North. This project aims to develop a residential area with floating, sustainable, circular houses. Initiated by a group of Amsterdam residents, it will be completed in 2020 and is run as a foundation.

Third, culture is used as a medium to convey the environmental sustainability message to a wider audience. The ADC aims to involve and inspire the public to join 'the social transition to a contemporary circular lifestyle' (De Ceuvel 2019). This is based on the belief that this transition is both technical and cultural (Ibid.). It occurs through the variety of cultural events and activities that regularly take place at De Ceuvel, with the aim of 'bringing awareness, making knowledge publicly accessible (...) making [environmental] sustainability low threshold and fun' (Interviewee 1, assistant to a co-founder, 2019). De Ceuvel has a goal to reach out to a broader audience, from the already 'converted' to the multi-ethnic inhabitants of the neighbourhood; Amsterdammers who have never crossed the River IJ; and children and tourists.[12]

However, most of the cultural activities do not reach the local neighbourhood, a relatively poor area. Indeed, the audience attracted is mostly made up of those 'who can associate [themselves] with this place [...] All these people who've been gentrifying the hell out of these neighbourhoods, they like coming here' (Interviewee 2, tenant, 2015). The locals do not go to De Ceuvel, even though some attempts have been made to reach out to them, for example with a traditional Dutch music festival and information sessions. It may be that the locals regard the café and activities as too expensive and view De Ceuvel as filled with 'strange people'. Meanwhile, those at De Ceuvel may develop a more snobbish and colonizing attitude towards the neighbourhood. In this way, the two parties become separate islands.

In conclusion, according to the interviewees, working at De Ceuvel enables them to realize their shared values through their professional work, as well in terms of interactions with the other tenants. However, the only public involved are the 'already converted' rather than people in the local neighbourhood. Our findings are confirmed by a recent report on value creation produced by trans-sectoral cooperation and commissioned by the Dutch Ministry of the Interior and Kingdom Relations. One of the tenants said that 'what's mentioned as culture in De Ceuvel is limited to an environmental communication tool for mainly white millennials—very Calvinist. There's little connection, little inter-generational, little diversity in De Ceuvel. If I hadn't invested 25K euros in my boat, then I'd probably already have left the place' (Weiler 2019, p. 65; authors' translation).

8.6 Conclusions

This chapter aimed to improve understanding of the relationship between culture and sustainable development using the case of De Ceuvel, an experiment where culture and environmental sustainability are intertwined. The goal was to assess

[12] De Ceuvel organizes art workshops for young children, as well as events like Ceuvel Kids, with nature-exploring activities or Easter egg hunts in and around the café. It organizes public tours for adults every Sunday. De Ceuvel is also part of the Public Art tour in Amsterdam North, and is one of the options available as a place to visit during the 'We Make the City' place-making conference.

why and how cultural and creative entrepreneurs commit to sustainable development. Since its inception, De Ceuvel has attracted creative entrepreneurs who wanted to work on environmental challenges and use culture to make the wider public aware of these issues and possible solutions. The decisive factors mentioned by our interviewees were: cheap rents; opportunities to collaborate and network with like-minded creative ecopreneurs; the amenities available (the café, the aesthetics, the nature and cultural activities); and the major technological innovations in relation to environmental sustainability. However, not everything in the garden is rosy. Indeed, our analysis suggests that funding opportunities have shaped the kinds of activities engaging our artists or entrepreneurs, with the majority following the money and possibilities that the circular economy may offer.

Along with a temporality paradox, another version may emerge. De Ceuvel has become an environmentally sustainable playground, but it is also part of a larger redevelopment process involving the entire Buiksloterham area. The broader plan is to expand the gentrification process already occurring in Amsterdam North. Indeed, while over-tourism characterizes central Amsterdam, Amsterdam North is seeing a rise in new residential areas and has a booming night-time economy, with new cafés and restaurants. A recently opened metro line means the area is only a few minutes from the city centre. Nevertheless, will the creative city saga continue, camouflaged as a sustainable or circular city, with the displacement of residents and precarious creative labour? Affordable spaces for artists in Amsterdam have become a mirage, while Amsterdam North, once a haven for alternative artists' communities (for example, NDSM Art City) may lose its final members, as already experienced by the rest of Amsterdam. Indeed, many Amsterdam-based artists are relocating to the closest city of Haarlem or are flooding to the rising creative city of Rotterdam.

The redevelopment of De Ceuvel can be assessed as a success in terms of both how it has advanced environmental technologies and some of the long-term effects it will have in relation to future generations. However, when cultural and social sustainability are considered, it is not possible to conclude that intragenerational equity is ensured. A fair distribution of economic and social opportunities and resources, along with fair access to cultural assets for disadvantaged groups of the present generation, were not part of the project. There is also a lack of recognition of the interdependence among its economic, environmental, social and cultural processes. An imbalance within De Ceuvel may also emerge in relation to the distribution of resources, the time invested and the rewards for different tenants. The entrepreneurs and organizations that have benefitted the most financially and reputationally may indeed be among the circle of founders, namely the architects and project managers who won the tender. For them, De Ceuvel may have been like a business card for future redevelopment projects that want environmental sustainability and to build a community of like-minded people. De Ceuvel may then be seen as a free publicity tool that increased the reputation, visibility and media attention paid to those involved, with skyrocketing commissions, awards and building projects being the result. One might wonder about the extent to which the type of creative entrepreneurs attracted—mostly in the architecture and design sectors—can explain how De Ceuvel was conceived and further developed.

In conclusion, De Ceuvel cannot be regarded in terms of either Throsby's (1993) conception of a culturally sustainable development project or as a sustainable creative area discussed by Kirchberg and Kagan (2013). The role of culture has mostly been instrumental in bridging other, mainly financial and environmental, goals. The empowerment of diverse society groups is minimal or lacking. A tender that would have also included social and cultural sustainability, combined with a performance assessment of its mid-term and final results, may have helped to settle a more inclusive project from the start. While environmental awareness may have been raised among the 'already converted', it is now time to experiment with a systemic form of project development that includes cultural and social sustainability.

References

Andres, L., and B. Grésillon. 2013. Cultural brownfields in European cities: A new mainstream object for cultural and urban policies. *Inernational Journal of Cultural Policy* 19 (1): 1–23.

Banks, M. 2017. *Creative justice: Cultural industries, work and inequality*. London: Rowman & Littlefield.

Banks, M., and J. O'Connor. 2009. After the creative industries. *Inernational Journal of Cultural Policy* 15 (4): 365–373.

Braun, E., and M. Lavanga. 2007. *An international comparative Quickscan into national policies for creative industries*. Rotterdam: Euricur for the Ministry of Education, Culture and Science of the Netherlands.

De Ceuvel. (2019). *De Ceuvel website*. Retrieved 22 Apr, 2019, from http://deceuvel.nl/en/

De Geus, M. 1999. *Ecological utopias. Envisioning the sustainable society*. Utrecht: International Books.

Diesendorf, M. 2000. Sustainability and sustainable development. In *Sustainability: The corporate challenge of the 21st century*, ed. D. Dunphy, J. Benveniste, A. Griffiths, and P. Sutton, 19–37. Sydney: Allen & Unwin.

Duxbury, N., A. Kangas, and C. De Beukelaer. 2017. Cultural policies for sustainable development: Four strategic paths. *Inernational Journal of Cultural Policy* 23 (2): 214–230.

Evans, G. 2009. Creative cities, creative spaces and urban policy. *Urban Studies* 46 (5-6): 1003–1040.

Foord, J. 2009. Strategies for creative industries: An international review. *Creative Indian Journal* 1 (2): 91–113.

Hawkes, J. (2001). The fourth pillar of sustainability: Culture's essential role in public planning. *Cultural Development Network*. Retrieved 10 Mar, 2015, from http://www.culturaldevelopment.net.au/community/Downloads/HawkesJon(2001)TheFourthPillarOfSustainability.pdf

Hesmondhalgh, D., M. Nisbett, K. Oakley, and D. Lee. 2014. Were new labour's cultural policies neoliberal? *Inernational Journal of Cultural Policy* 21 (1): 97–114.

Kagan, S., and J. Hahn. 2011. Creative cities and (un)sustainability: From creative class to sustainable creative cities. *Culture Local Governance* 3 (1): 11–27.

Kirchberg, V., and S. Kagan. 2013. The roles of artists in the emergence of creative sustainable cities: Theoretical clues and empirical illustrations. *City, Culture and Society* 4 (3): 137–152.

Klamer, A. 2002. Accounting for social and cultural values. *De Economist* 150 (4): 453–473.

———. 2017. *Doing the right thing: A value based economy*. London: Ubiquity Press.

Lavanga, M. 2004. Creative industries, cultural quarters and urban development: The case studies of Rotterdam and Milan. In *Urban management in Europe vol. II: Towards a sustainable development*, ed. G. Mingardo and M. van Hoek, 121–157. Rotterdam: Erasmus University Rotterdam, MEMR - The Thesis Series.

———. (2006). *The Contribution of cultural and creative industries to a more sustainable urban development: The case studies of Rotterdam and Tampere*. In Proceedings of the 14th International Conference of the ACEI (Association of Cultural Economics International), Vienna

———. 2009. Becoming a sustainable creative city. Cultural industries and urban (re)development in Tampere. In *Culture and the city*, ed. F. Eckardt and L. Nyström, 217–239. Berlin: Berliner Wissenschafts-Verlag.

———. 2013. Artists in urban regeneration processes: Use and abuse? *Territoire en Mouvement* 17: 6–19.

———. 2020. Cultural districts. In *Handbook of cultural economics*, ed. R. Towse and T. Navarrete, 3rd ed., 174–182. Cheltenham: Edward Elgar Publishing.

Levitas, R. 2000. For Utopia: The (limits of the) utopian function in late capitalist society. *Critical Review of International Social and Political Philosophy* 3 (2–3): 25–43.

Metabolic, Studioninedots and DELVA Landscape Architects. (2014). *Circular Buiksloterham. Transitioning Amsterdam to a circular city. Vision and ambition. Executive summary.* Amsterdam

Morton, J. (2018). *TEDxAmsterdam takes a tour of urban innovation playground De Ceuvel*. TEDx Amsterdam. Retrieved 22 Apr, 2019, from https://tedx.amsterdam/explore/message/tedxamsterdam-takes-a-tour-of-urban-innovation-playground-de-ceuvel

Oakley, K. 2004. Not so cool Britannia: The role of the creative industries in economic development. *International Journal of Cultural Studies* 7 (1): 67–77.

Oakley, K., and J. O'Connor. 2015. Culture and the city. In *Routledge companion to the cultural industries*, ed. K. Oakley and J. O'Connor, 201–211. London: Routledge.

Oakley, K., and J. Ward. 2018. The art of the good life: Culture and sustainable prosperity. *Cultural Trends* 27 (1): 4–17.

Pratt, A. 2008. Creative cities: The cultural industries and the creative class. *Geografiska Annaler* 90 (2): 107–117.

Rozentale, I., and M. Lavanga. 2014. The 'universal' characteristics of creative industries revisited: The case of Riga. *City, Culture and Society* 5 (2): 55–64.

Sargent, L.T. 2005. What is a Utopia? *Morus* 2: 153–160.

Sargisson, L. 2000. Green utopias of self and other. *Critical Review of International Social and Political Philosophy* 3 (2–3): 140–156.

Throsby, D. 1995. Culture, economics and sustainability. *Journal of Cultural Economics* 19 (3): 199–206.

———. 2008. Linking cultural and ecological sustainability. *International Journal of Diversity Organisations Commun Nations* 8 (1): 50–53.

———. 2017. Culturally sustainable development: Theoretical concept or practical policy instrument? *Inernational Journal of Cultural Policy* 23 (2): 133–147.

United Nations. (2019). *Sustainable development goals*. Retrieved 22 Apr, 2019, from https://sustainabledevelopment.un.org/?menu=1300

United Nations Department of Economic and Social Affairs. (2018). 68% of the world population projected to live in urban areas by 2050, says UN. *UNDESA News*. Retrieved 27 Jan, 2020, from https://www.un.org/development/desa/en/news/population/2018-revision-of-world-urbanization-prospects.html

United Nations World Summit. (2005). 2005 World summit outcome, Resolution A/60/1. *UNAIDS*. Retrieved 22 Apr, 2019, from http://data.unaids.org/Topics/UniversalAccess/worldsummitoutcome_resolution_24oct2005_en.pdf

van der Borg, J., A.P. Russo, M. Lavanga, and G. Mingardo. 2005. *The impacts of culture on the economic development of cities*. Rotterdam: Euricur Report.

Vanolo, A. 2013. Alternative capitalism and creative economy: The case of Christiania. *International Journal of Urban and Regional Research* 37 (5): 1785–1798.

WCED. 1987. *Our common future*. Oxford: Oxford University Press.

Weiler, C. 2019. *A study into the contribution of trans-sectoral cooperation in informal area development to the creation of social value*. The Hague: Atelier Rijksbouwmeester for the Ministry of the Interior and Kingdom Relations of The Netherlands.

Zukin, S. 1989. *Loft living: Culture and capital in urban change*. New Brunswick: Rutgers University Press.

Chapter 9
Creative Labour, Metabolic Rift and the Crisis of Social Reproduction

Frederick Harry Pitts

9.1 Introduction

This chapter investigates how creative labour—particularly in creative industries like graphic design, advertising and branding—connects with two appearances of crisis in the contemporary age. On the one hand is a contemporary crisis of social reproduction—a breakdown in the relationship between work, the wage and the capacity to consume the goods and services sufficient to subsist (Caffentzis 2002; Fraser 2016; Leonard and Fraser 2016). On the other is the appearance of this 'global organic crisis' (Bakker and Gill 2003; Gill 2016) as a crisis in the human metabolism with nature, the kind of ecological cataclysm that Marx conceptualizes as a 'metabolic rift' (Foster 1999, 2016; cf. Moore 2011, 2017). From this perspective, human labour is the mediation of the human metabolism with nature, a metabolism that is synonymous in capitalist society with subsistence by means of the reproduction of human life as labour power in a social, biological and, crucially, ecological sense. This chapter focuses on how the crisis of social reproduction, synonymous with that of the ecology itself (Fraser 2016), is expressed in a crisis of the social forms of mediation in which the antagonisms and contradictions that criss-cross this conflicted reproduction are expressed and managed, among them the state, value, money and the commodity form itself (Dinerstein and Pitts 2018; Lombardozzi and Pitts 2019).

From a wider perspective the constraints on the human capacity to reproduce life as labour power centre less on a sudden crisis or rift with the metabolism with nature than they do on a more permanent contradiction that cuts to the core of capitalist society itself (Lombardozzi and Pitts 2019). This relates, as we shall see, to the tension between production and consumption in a society based in the separation of human subjects from nature and the independent collective or individual means to

F. H. Pitts (✉)
University of Bristol School of Management, Bristol, UK
e-mail: fh.pitts@bristol.ac.uk

K. Oakley, M. Banks (eds.), *Cultural Industries and the Environmental Crisis*,
https://doi.org/10.1007/978-3-030-49384-4_9

secure the means of living. This chapter explores interventions and contradictions at the level of the workplace, the market and the state with reference to the creative industries and the pivotal position they assume in the production, consumption and circulation of goods and services in capitalist society. The chapter ends by considering the policy and practical actions taken by state and autonomous actors to ensure the continuity and, crucially, *sustainability* of creative labour against a backdrop of constrained spending, economic uncertainty and sectoral competition.

My argument is informed by Open Marxism (Bonefeld et al. 1992), the New Reading of Marx (Bellofiore and Riva 2015), Social Reproduction Theory (Bhattacharya 2017) and Metabolic Rift Theory (Foster 1999, 2016). I use Marx's critique of political economy to capture the character and contradictions of social forms of state, value, money and the movement of commodities (Bonefeld 2014). My argument is based on observations and propositions derived from empirical research projects spanning over 7 years which investigated creative agencies and their employees, creative freelancers, and some of the organizations that have sprung up recently to support and represent freelancers. I reflect on the experiences and testimonies gathered in this ethnographic and interview-based research on 'cultural intermediaries' (Kuipers 2014).[1]

The chapter consists of three sections. First, it offers an account of the social metabolism and its association with the metabolic rift between humans and nature and the crises of overproduction and social reproduction the concept describes. Second, it shows how this impacts upon the creative industries and the work performed within them and the steps taken by the state and other actors to mediate and mitigate the antagonisms and contradictions that arise. Third, a concluding section argues that autonomous projects to reimage and recreate creative work may have the greatest potential to withstand and challenge socio-ecological crisis.

9.2 Metabolic Rift, Social Reproduction and Marx's Central Contradiction

Materially estranged from the 'natural conditions of their existence', humans must nonetheless dominate and objectify these conditions in order to live (Foster 1999, p. 383) even as their transformation to serve human needs and purposes erodes and degrades these conditions. For Marx, the metabolism is a fixed feature of the human labour process not only under capital but in general; production always formally

[1] These projects are as follows: The Social Validation of Abstract Labour-Time: A Case Study of Billable Hours in the Creative Industries in the UK and the Netherlands (2011–2015, ESRC 1+3 PhD Studentship); E-Rhythms in Freelance Creative Work (2014, EU COST Action IS1202 Dynamics of Virtual Work Short-Term Scientific Mission Funding); Organizing Self-Employed Workers in the Creative Industries (2017–2018, ESRC Impact Acceleration Award). The relevant outputs from these projects are Pitts (2015a, b, c, d, 2016a, b, 2017a, b, 2019). To avoid repetition individual citations are not given each time a finding from these outputs is discussed.

mediates what Marx calls the 'separation' of humankind from nature (Marx 1973, p. 489; see also Foster 2016, p. 403; Meszaros 1970). Thus the 'social metabolism' (Foster 2016, p. 403–404), or 'social-ecological metabolism' (Foster 1999, p. 380), mediated in capitalist society by the expenditure of labour-power and exchange of commodities, is a particular appearance of the 'universal metabolism of nature' that governs the transformation of nature into objects of human need.

The separation from nature necessarily implies that human life is never immediate and always-already mediated. For Marx, what distinguishes 'the worst of architects from the best of bees' is a separateness of humankind from nature encapsulated in the capacity of the former to conceive of designs and enact them in products in which both external nature is transformed into something useful and humans realize their subjectivity as such (Marx 1976, p. 284–285). These products—in the shape of commodities, money, the state and so on—then take on a life of their own, standing over their producers whose lives and subsistence they mediate in turn. Humans transform nature but under social and historical conditions not of their choosing and with different and sometimes unintended consequences (Foster 1999, p. 390).

The social metabolism under capitalism represents only the current 'particular, alienated form' of the metabolic relationship between humans and nature (Foster 2016, p. 404). This alienation, however, stems from something intrinsic to the human condition and cannot be easily wished away.[2] The mediated character of the human metabolism takes on different guises in different times and places. Labour is a form of mediation assumed by the 'metabolism between man and nature' (Marx 1976, p. 133; see also Postone 1993), the incommensurability of its products is mediated in the social form of money (Sohn-Rethel 1978), and the antagonisms and contradictions that arise around the capital-labour relationship in capitalist society and the constrained capacity of workers to consume the products they produce are mediated in the social form of the state (Holloway and Picciotto 1977). The metabolism between humans and nature, then, is the basic process these forms of mediation express and to which they reduce, constituting 'the basis on which life is sustained and growth and reproduction becomes possible' (Foster 1999, p. 383). But the concept captures not a static or essentialist notion of nature and society but a 'highly dynamic relationship reflecting changes in the ways human beings mediate … between nature and society through production' (Foster 1999, p. 388).

The concept of metabolism has been used to theorize environmental crisis based upon the notion of a 'metabolic rift' (Foster 1999) which, for Marx, is 'an irreparable rift in the interdependent process of the social metabolism' (Marx 1981, p. 949–950). The degradation of natural resources by industry is synonymous with the degradation of labour-power, which constitutes the mode of existence through which humans subsist under capital, it being, as Marx writes, 'the material of nature transposed into a human organism' (Marx 1976, p. 323). Both sides, natural and human, are exhausted in the 'metabolic interaction between man and the earth'

[2] For a more detailed discussion of Marx's changing conceptualization of these issues, see Pitts et al. (2019).

(Marx 1976, p. 637–638). Whilst the rift represents a contradiction in capitalism itself, contemporary conditions have exacerbated the environmental degradation in which it results. However, as Moore (2011, p. 2) suggests, the temporality with which claims of a metabolic rift are endowed does not quite capture how the crisis and the antagonisms and contradictions on which it centres are not new or novel but rather permanent and constitutive of capitalism *tout court*, synonymous with what, for Marx (1973, p. 748), represented its basic crisis tendency: the overproduction of goods relative to the capacity of workers to consume them.

Moore (2011, p. 12) argues that the metabolic rift is best understood in terms of the relationship between the overproduction of goods, subject to the constrained conditions of social reproduction associated with the 'capital-labour antagonism', and the 'capital-ecology antagonism' that serves to further erode and degrade the 'conditions of production' themselves, fettering accumulation and 'reinforc[ing] difficulties the system already faces in the realization of surplus value through the sale of commodities.' This rift is expressed just as much in 'defunding public education or the deterioration of vital infrastructures' as it is in 'soil exhaustion or deforestation', in all cases exhibiting the central contradiction whereby capitalism exhausts that on which it relies—'human labour power, built environments and resources' (Ibid., p. 14)—increasing barriers to accumulation and reproduction but with no promise of terminal collapse or resolution.

9.3 Overproduction, the Creative Labour Process and State Intervention

This rift carries implications for the position of the creative industries in capitalist society. Creative industries play a pivotal and not merely peripheral role in the production, circulation and consumption of goods and services as commodities (Nitzan and Bichler 2009). In endowing goods and services with meaning—via advertisements, packaging, websites and even intervention in the design and production of goods themselves—creative industries like graphic design, branding and advertising play a vital part in the commensuration of commodities in the market (Lotz 2014; Williams 2005). In so doing they address a contradiction at the centre of the global economy: the spending power of consumers is always limited by its relation to a wage that must, for profit's sake, be less than the value of the goods and services produced and consumed by those in receipt of it (Heinrich 2013; Marx 1976). The conditions for the consumption of commodities are constrained by the conditions of their production. The global economy, therefore, must constantly work against the underlying tendency to crises of overproduction (Clarke 1992), symptomatic of a system so rapacious that it risks environmental ruin (Malm 2018). Creative industries like graphic design, advertising and branding intervene in this context by aiding the circulation of goods and services as commodities, endowing them with desirability and meaning even where the material conditions for their consumption militate against it.

This contradictory context renders creative industries themselves subject to new contradictions. As they intervene to mould consumer sentiment on behalf of their clients, their success comes to hinge on its vagaries, the capacity to consume always circumscribed by the constraints established in antagonistic social relations of production beyond the influence of creative agencies and their clients to change. The value of commodities is realized in their exchange; monetary value is socially validated and not implied in any material or objective property of the thing itself (see Heinrich 2012). Agencies, which depend on the reproduction and expansion of a capitalist system that dominates them, live or die by the validation of the goods and services they help sell and circulate (Bonefeld 2014). The relationship with the client introduces new ways of performing, measuring and evaluating work into the creative spaces of agencies (Moor and Julier 2009).

In this context, creative work is validated by the client. In a society where the value of human activity is recognized monetarily through the sphere of waged work as a condition of its reproduction, creatives can be creative only within the criteria of a value-generating corporate outcome (Hesmondhalgh and Baker 2011). The 'ordinary' capacity to be creative in an artistic or visual sense in capitalist society can be objectified only in certain kinds of commodified goods and services (Williams 1965), leading some commentators to bemoan the situation whereby those most capable of producing aesthetic beauty and satisfaction are engaged in commercial spheres that squeeze all that is good from that capacity for ends considered bad (Baran and Sweezy 2013). Whilst, as some scholars have pointed out, there is unexpended political leverage when creatives are responsible for populating the consumer imaginary and ensuring the circulation of goods in capitalist society, there is little prospect presently of this being translated into anything like the revolutionary potential some have seen as latent in this state of affairs, whereby 'the elimination of advertising as we know it today would require the elimination of capitalism' (Baran and Sweezy 2013; Foster and McChesney 2013).

The underlying contradictions in capitalism itself make the work that creatives perform uncertain and insecure. Their continued ability to be creative depends on an income linked to the fortunes of clients operating in a crisis-ridden global economy undermined by the limitations placed on spending by antagonistic social relations of production, and in which creative goods and services possess an inherent unknowability (Caves 2002). Creative industries like design, branding and advertising might help combat the constraints placed on consumption by this central contradiction (Baran and Sweezy 2013), but, owing to the persistence of antagonistic social relations of production and unequal distribution, always tread a thin line between securing an upturn in a client's sales and falling victim to the contradiction themselves when circulation seizes up in times of crisis. This uncertainty and insecurity renders work precarious, and influences the spread of non-standard contracts of employment—including the use of freelancers—that allow flexibility in response to market conditions (Gill and Pratt 2008; Hesmondhalgh and Baker 2008; Ross 2008). This shows that the social metabolism, and the metabolic rift, are not solely abstract, macro-level processes remote from the experience and practice of work. Aside from the effects on the contractual form under which creative labour proceeds, the dynamics of the metabolism play out on an everyday basis by influencing the decisions made by organizations about how to manage the creative labour process, and impact upon the performance of creative work itself.

Under these circumstances the experience of creative labour becomes structured by the external context in which it operates, rather than anything intrinsic to the performance of that labour itself. The condition for the ability of creatives and artists working in these fields to pursue their vocation is their service to clients in addressing the contradictions of commodity consumption. Their work becomes structured, evaluated and measured in line with this relationship to clients in other corporate sectors—contrary to the claims made in the wake of post-operaist theories of 'immaterial labour' that creative work is somehow immeasurable (Boehm and Land 2009; Lazzarato 1996). Indeed, in creative agencies work *is* measured, through an 'agency time' (Ladner 2009) constructed on the basis of the hours billed out to clients (Dorland 2009; for case studies of billables in other sectors see Alvehus and Spicer 2012; Ross 2003; Yakura 2001), a construction not of how long a design, brand or concept will take but rather of the budgetary constraints clients bring to the negotiating table. Some agencies seek to develop, with varying success, alternative forms of measure, associated not with the hour but with the end result, for instance, or with a demonstrable impact on sales or website visitors where data is available. The development of these alternatives centres on a conflict between agency and client over control of the creative labour process. So beholden are agencies to their clients for their continued reproduction that a project is considered successful if the client is suitably impressed to return with future custom, even where, in securing this custom, agencies have pitched for work writing in an eventual loss.

This has an impact on the experience and practice of creative work. For instance, if a job is priced at a loss to attract a client, the work in meeting the brief (see Dorland 2009) will be all the more intense and unconfined by the normal structure of the working day. These factors do not only make the experience of creative work alienating and exploitative for individual creatives but also make the management and day-to-day running of creative agencies subservient to principles other than those intended by their founders, partners and owners. The implementation of new philosophies and technologies of project management, such as Agile working methods (Hodgson and Briand 2013), are one means by which control of the creative labour process can be wrested back from the client, so that the agency's work can become more committed to making and creating.

The disjuncture between the expectations and reality of creative work centres on the conflict between the desire to be creative and the compulsion to render that creativity useful, understandable and valuable to an external context. A worker's desire to be creative must submit to the time discipline of the billable hour: work is simplified, standardized and deskilled, so that hours can be compared like-for-like, and their contents more easily measured. This reduction of their labour in practice is accelerated by certain imperatives linked to the corporate context in which agencies find themselves. Despite depending on the stimulation of feelings of novelty for their success, the livelihoods of agencies and their employees are, paradoxically, reliant on repeat business and long-term accounts with clients. This longevity renders the work that goes into a given design slighter in character and complexity each time a client contracts the agency for a new piece of work carrying their brand identity. This replicates the tendency towards rationalization and standardization inherent in many creative industries (Ryan 1992), in spite of the apparent resistance of creative processes and products to it (Hesmondhalgh and Baker 2011). Standardization of products, identities and advertising by corporate and state regulation standardizes the work in turn. Through mergers and acquisitions, large conglomerates subsume smaller firms and standardize their range of brand identities internally. As agencies seek repeat business from smaller client rosters, in competition with other agencies seeking the same claim on whatever small corner of the market they can secure, these processes of standardization gather pace.

The pressure placed on creative labour and the practices of creative industries forces some workers to leave agencies in revolt against the intensification of work regimes and to become freelancers. This represents a wider fragmentation of the creative field that can be related to the structural conditions in which it operates. It can result in an 'hourglass' shape in the sector, whereby large advertising and media conglomerates absorb medium-sized agencies through acquisition, who in turn compete against smaller upstart agencies set up by creatives dissatisfied with the creative freedom afforded them in their old jobs, or freelancers picking up pieces of work from clients directly—a subset marked at the bottom end by small-time bedroom designers recruited for tasks farmed out by companies using new online platforms. From below, this groundswell of independent work and upstart studios places extra pressures on a top end of the market already locked in fierce competition for the available clients.

There is an interplay between these dynamics and the process of creative clustering (Hartley et al. 2013, p. 17–20; Hesmondhalgh 2013, p. 170–174) in which local authorities (for example, Amsterdam Economic Board in the Netherlands) and national-level organizations for the creative industries (for example, Creative England) place extensive resources to stimulate their formation and maintenance. The clustering of creatives and agencies in certain locales functions through a contradictory momentum that the state, locally and nationally, is called upon to superintend and contain. These interventions can intensify the effect the geography of the creative industries has on those subject to the competitive struggle produced by the progressive fragmentation and reformation of the sector. To begin with agencies move into hip urban areas, following where artists and creatives lead in search of cheap rent and studio space (Florida 2002). Following this influx of creatives, the gentrification of formerly cheap areas means they become unaffordable both to the young workforces on which the agencies rely and—at the level of ground rent, business rates and so on—to the agencies themselves. Higher overheads become increasingly unsustainable as clients spend less through belt-tightening or agencies undercut themselves and others in pursuit of repeat work. In the context of this situation—where the creative sector fragments in personnel and sectoral composition, and becomes subject to competing interests at the sectoral level and in individual workplaces—there are notable attempts by the state to smooth over the contradictions to which the sector is subject (see Christophers 2007; Hesmondhalgh et al. 2015; Hesmondhalgh and Pratt 2005; O'Brien and Oakley 2017), so as to protect and guarantee the pivotal position creative industries take in wider schemas and circuits of consumption and the realization of profit and value.

But as we have seen, the underlying contradictions run deeper than is possible for the state as a form of mediation alone to mitigate. The pressures that compromise what makes creative industries what they are to begin with—creativity (Galloway and Dunlop 2007)—must be managed carefully, and the sector must be made to cohere so as to perform its function within the economy as a whole as effectively as possible. Examples include the long sequence of cultural and creative industries policy begun first under the Greater London Authority in the 1980s and then continued forcefully under New Labour (Hesmondhalgh and Pratt 2005; Hesmondhalgh et al. 2015). The inheritance is clear in organisations like NESTA, who have launched large-scale attempts to measure the value of the creative industries and perfect industrial classificatory schema to better capture the points at which they intervene in the economy as a whole.

The formation of so-called 'quangos' like Creative England that supervise the clustering of creative firms at a local and national level, and the direct interest taken by local organisations like the Amsterdam Economic Board in keeping a steady flow of new creative labour coming to the city to populate its proliferating ecology of smaller and larger agencies, also exemplify the various ways in which the state develops its own means to manage creative labour's imbrication in the contradictions that result from its pivotal position in the circulation of goods and services as commodities in the market. These measures institutionalize creative industries within the apparatus of the state in its superintendence of capital, economy and competition between companies.

These forms of state intervention seek to neutralize the contradictions over which agencies, creatives and clients contend. These same contradictions drive agencies to develop new ways of measuring work and new modes of project management that wrest control from clients. They drive freelancers to leave agencies and go it alone, grouping together with others in innovative collectives of individuals. They compel local authorities and quangos to step in and supervise the haphazard and ill-fated clustering of creatives and agencies in urban silos. All this seeks the impossible resolution of contradictions so that business can continue as usual, offering no alternative to the impasse.

9.4 Conclusion: Where Hope Dies, Action Begins

This chapter has suggested that rather than posing a way out, creative industries policy in the UK is related to managing the chaotic reproduction of a sector in the face of a series of contingent factors: economic crisis and a contraction in consumer spending on goods and services; an unsustainable cycle of gentrification in urban areas; and a sectoral fragmentation as big media and advertising conglomerates subsume small and medium firms, from which disenchanted creatives in turn break away to go freelance or establish upstart small studios. Increased competition then follows from below, as clients tighten budgets and agencies price themselves lower to secure continued custom against the competition. All these factors relate to the irreparable rift in the social metabolism and the permanent crisis of social reproduction against which creative industries must struggle at the coalface of the exchange and circulation of the commodities capitalism habitually overproduces in the context of this constrained material basis.

State responses aim to protect the pivotal position played by creative industries in the contemporary economy against this antagonistic undertow. Superintending the sustainable clustering of creative firms, advertising cities and countries as great places to be creative, and mediating the unplanned competition of isolated companies, they seek the temporary resolution of contradictions so that business can continue as usual, rather than confronting their imbrication in a crisis of social reproduction by supporting the development of alternative ways of being creative whilst also possessing the certainty of securing the conditions of life. These measures secure the continuity of creative industries but resolve only temporarily the contradictory status to which they are subject, concerned with creativity but beholden to a corporate context that imports economic uncertainty and more intense, exploitative working conditions.

The intractability of these contradictions renders it difficult to envisage policy responses capable of overcoming or even temporarily suspending them, partly owing to the state's own status as their necessary form of mediation. Just as with the so-called crisis of social reproduction (Lombardozzi and Pitts 2019), the metabolic rift does not describe a temporary disruption resolvable in, and through, the social relations and social forms specific to capitalist society, or perhaps any society. It represents instead a much more permanent but still dynamic state of affairs, logically and historically hardwired into capitalism's contradictory and conflict-ridden functioning. Moore (2011, p. 10) is therefore correct to note the 'promiscuity' with which crises are ascribed to contemporary capitalism. What masquerade as unhappy consequences—'demographic collapse, soil exhaustion', to which we might add in-work poverty and lack of access to basic subsistence goods—are in fact constitutive of the purported cause—the mediation under capitalism of an alienation intrinsic to what it is to be human.

This indicates a wider issue regarding the social metabolism and the rift to which it is subject. For some, like first-generation Frankfurt School critical theorists (Adorno and Horkheimer 1972; Schmidt 1971), the separation of humans from nature constructs a 'Weberian-like iron cage from which there [is] no escape' (Foster 2016, p. 395). As Moore (2017, p. 295) argues, this rests on a pessimistic characterization of humankind as in some way fallen from a state of Edenic grace. In posing no escape from this state of fallenness, this strand of social theorizing rules out the possibility of a return to an unmediated and prelapsarian life of oneness with nature, permitting only the choice of better and worse mediations of our alienated relationship with the world around us. In this it provides a platform for navigating the impasse in the same spirit captured in a placard at the recent

Extinction Rebellion protests in London: 'where hope dies, action begins'. Along these lines, the limitations placed upon the human ability to overcome the metabolic rift may liberate subjects to create new mediations of the metabolism rather than stand paralysed in fear of a lack of feasible systemic alternatives. As Marx (1981, p. 959) writes:

> Freedom [...] can consist only in this, that socialized man, the associated producers, govern the human metabolism with nature in a rational way, bringing it under their collective control instead of being dominated by it as a blind power; accomplishing it with the least expenditure of energy and in conditions most worthy and appropriate for their human nature.

But, Marx stresses, it remains a realm of necessity nonetheless. Accepting this limitation saves effort otherwise wasted attempting the incomplete resolution of irresolvable contradictions. We need to reimagine what creative labour is and can be within—and not outside—the confines of its current commercial context and construct new institutions that are autonomous from the state and capable of reproducing the conditions for more emancipated work in the creative industries. Indeed, creatives are already organizing autonomously of the state. At the individual level at least, the institutionalization of the autonomy made possible by freelancing represents one such alternative for creatives who leave agencies because they feel their creativity has been stifled by the corporate client relationship. This dynamic sits at the meeting point of two tendencies. On one hand, what we have termed a crisis of social reproduction constrains consumer spending and limits the capacity of agencies to be creative, insofar as they depend, for the opportunity to do so, on their recruitment by clients to help sell commodities that now meet a declining customer base. On the other, freelancers, free of the compulsion to foreshorten their creativity in the corporate context of that crisis, become subject to the very broken link between work, the wage and the means of living on which the same crisis centres. With the freedom of freelancing comes precariousness and insecurity of social reproduction, for which none of the statutory supports enjoyed by formal employees are available.

To address this, freelance creatives have led the way in constructing institutionalizations of other ways of working and living creatively. These include new forms of trade union such as those among freelancers in North America (Cohen 2012, 2016); Business and Employment Cooperatives like the French *Clara Cooperative Culturelle* (Conaty et al. 2015); mutual assistance schemes like the Dutch Broodfonds into which freelancers contribute monthly sums to guarantee payment for members sick or out of work; and guild-style networks based on the struggle to establish emergent forms of professional identity, such as those found in the Italian experience (Boffo 2014; Bologna 2018). Two particularly interesting examples are SMart, a salaried platform cooperative based in Belgium but with a presence in several European countries (Graceffa and de Heusch 2017; Xhauflair et al. 2018; Zanoni 2018, 2019), and Indycube, a co-working cooperative and trade union for the self-employed originating in South Wales and steadily extending its reach into the wider United Kingdom (Fuzi 2015). SMart intermediates between freelancers and their clients by employing the former and invoicing the latter, guaranteeing the freelancer all the rights, privileges and security of formal salaried employment whilst sustaining the autonomy and independence associated with freelance work. Indycube offers co-working space to reconstruct a sense of community in the atomized new world of work and has started offering invoice-chasing services and advice to freelancers through a new cooperative trade union—with a view to implementing the SMart model in the UK (Pitts 2019, Ross et al. 2017).

It must be noted that whilst they both owe their beginnings to creative milieus, neither is exclusive to those working in the creative industries alone. But, avoiding the temptation of an impossible escape, they instead promise to remediate creative labour so as to render it more sustainable and more resistant to the intractable contradiction between society's capacity to produce and its capacity to consume owing to the constrained ecological basis for both. Moreover their innovations speak not only to the creative sphere but to a wider audience of workers facing similar conditions. Creative labour, specifically at the freelance end, in many ways presents a limit case for contemporary work in general (Stahl 2014). The project-based precariousness of the work, the non-standard contractual statuses, the lack of traditional support networks in trade unions or social security: all these features, present elsewhere in the economy, take on exemplary and sometimes extreme forms in the creative industries. However, creative workers command a pivotal position in the circulation of goods and services (Baran and Sweezy 2013), and thus in the economy as a whole, not least in times of crisis. How creative workers individually and collectively leverage this pivotal position in the development of alternative ways of living and working (see Durrer et al. 2014; Elzenbaumer 2014) could therefore be influential for policy and political responses to the crisis of social reproduction and the intractable contradictions it expresses more widely. They have the potential, as Foster and McChesney (2013) put it, to 'reestablish the relationship between the consumer and producer of cultural work by undermining the estrangement from human needs and capacities enforced [in] bourgeois society', with all the socio-ecological consequences this implies.

References

Adorno, T., and M. Horkheimer. 1972. *Dialectic of englightenment*. London: Verso.

Alvehus, J., and A. Spicer. 2012. Financialization as a strategy of workplace control in professional service firms. *Critical Perspectives on Accounting* 23: 497–510.

Bakker, I., and S. Gill. 2003. *Power, production and social reproduction*. Basingstoke: Palgrave.

Baran, P.A., and P.M. Sweezy. 2013. Theses on advertising. *Monthly Review* 65 (3): 34–42.

Bellofiore, R., and T.R. Riva. 2015. The Neue Marx-Lektüre. Putting the critique of political economy back into the critique of society. *Radical Philosophy* 189: 24–36.

Bhattacharya, T., ed. 2017. *Social reproduction theory*. London: Pluto.

Boehm, S., and C. Land. 2009. No measure for culture? Value in the new economy. *Capital & Class* 97: 75–98.

Boffo, M. 2014. From post- to neo: Whither operaismo beyond Hardt and Negri? *Historical Materialism* 22 (3–4): 425–528.

Bologna, S. 2018. *The rise of the European self-employed workforce*. Milan: Mimesis International.
Bonefeld, W. 2014. *Critical theory and the critique of political economy: On subversion and negative reason*. London: Bloomsbury.
Bonefeld, W., R. Gunn, and K. Psychopedis. 1992. *Open marxism volume I: Dialectics and history*. London: Pluto Press.
Caffentzis, G. 2002. On the notion of a crisis of social reproduction: A theoretical review. *Commoner* 5: 1–22.
Caves, R. 2002. *Creative industries: Contracts between art and commerce*. Cambridge: Harvard University Press.
Christophers, B. 2007. Enframing creativity: Power, geographical knowledges and the media economy. *Transactions of the Institute of British Geographers* 32: 235–247.
Clarke, S. 1992. The global accumulation of capital and the periodisation of the capitalist state form. In *Open marxism volume I: Dialectics and history*, ed. W. Bonefeld, R. Gunn, and K. Psychopedis, 133–150. London: Pluto Press.
Cohen, N.S. 2012. Cultural work as a site of struggle: Freelancers and exploitation. *TripleC* 10 (2): 141–155.
———. 2016. *Writers' rights: Freelance journalism in a digital age*. Montreal: McGill-Queen's Press.
Conaty, P., Bird, A., Ross, C. (2015). *Not alone. Trade union and co-operative solutions for selfemployed workers*. Co-operatives UK. Retrieved 27 Feb, 2020, from https://www.uk.coop/sites/default/files/uploads/attachments/working-together_final_web-version.pdf
Dinerstein, A.C., and F.H. Pitts. 2018. From post-work to post-capitalism? Discussing the basic income and struggles for alternative forms of social reproduction. *Journal of Labor Society* 21: 471–491.
Dorland, A. 2009. Routinized labour in the graphic design studio. In *Design and creativity: Policy, management and practice*, ed. G. Julier and L. Moor, 105–121. Oxford: Berg.
Durrer, V., D. O'Brien, and D. 2014. Rethinking cultural intermediaries: Negotiating boundaries in the arts. In *The cultural intermediaries reader*, ed. J. Smith Maguire and J. Matthews, 124–138. London: Sage.
Elzenbaumer, B. (2014). *Designing economic cultures: Cultivating socially and politically engaged design practices against procedures of precarisation*. Ph.D. thesis, Goldsmiths, University of London
Florida, R. 2002. *The rise of the creative class*. New York: Basic Books.
Foster, J.B. 1999. Marx's theory of metabolic rift: Classical foundations for environmental sociology. *The American Journal of Sociology* 105 (2): 366–405.
———. 2016. Marxism in the anthropocene: Dialectical rifts on the left. *International Critical Thought* 6 (3): 393–421.
Foster, J.B., and R.W. McChesney. 2013. The cultural apparatus of monopoly capital. *Monthly Review* 65: 3.
Fraser, N. 2016. Contradictions of capital and care. *New Left Review* 100: 99–117.
Fuzi, A. 2015. Co-working spaces for promoting entrepreneurship in sparse regions: The case of South Wales. *Regional Studies, Regional Science* 2 (1): 462–469.
Galloway, S., and S. Dunlop. 2007. A critique of definitions of the cultural and creative industries in public policy. *International Journal of Cultural Policy* 13 (1): 17–31.
Gill, S. 2016. Critical global political economy and the global organic crisis. In *The Palgrave handbook of critical international political economy*, ed. A. Cafruny, S. Talani, and G.P. Martin, 29–48. Basingstoke: Palgrave Macmillan.
Gill, R., and A. Pratt. 2008. In the social factory? Immaterial labour, precariousness and cultural work. *Theory, Culture and Society* 25 (7–8): 1–30.
Graceffa, S., and S. de Heusch. 2017. Reinventing the world of work. *European Review of Labour and Research* 23 (3): 359–365.
Hartley, J., J. Potts, S. Cunningham, T. Flew, M. Keane, and J. Banks. 2013. *Key concepts in creative industries*. London: Sage.

Heinrich, M. 2012. *An introduction to the three volumes of Karl Marx's capital*. New York: Monthly Review Press.
———. 2013. Crisis theory, the law of the tendency of the profit rate to fall, and Marx's studies in the 1870s. *Monthly Review* 64: 15–32.
Hesmondhalgh, D. 2013. *The cultural industries*. 3rd ed. London: Sage.
Hesmondhalgh, D., and S. Baker. 2008. Creative work and emotional labour in the television industry. *Theory, Culture and Society* 25 (7-8): 97–118.
———. 2011. *Creative labour: Media work in three cultural industries*. London: Routledge.
Hesmondhalgh, D., and A.C. Pratt. 2005. Cultural industries and cultural policy. *International Journal of Cultural Policy* 11 (1): 1–13.
Hesmondhalgh, D., K. Oakley, D. Lee, and M. Nisbett. 2015. *Culture, economy and politics: The case of new labour*. Basingstoke: Springer.
Hodgson, D., and L. Briand. 2013. Controlling the uncontrollable: 'Agile' teams and illusions of autonomy in creative work. *Work, Employment and Society* 27 (2): 308–325.
Holloway, J., and S. Picciotto. 1977. Capital, crisis and the state. *Capital & Class* 2: 76–101.
Kuipers, G. 2014. Ethnographic research and cultural intermediaries. In *The cultural intermediaries reader*, ed. J. Smith Maguire and J. Matthews, 52–63. London: Sage.
Ladner, S. 2009. 'Agency time': A case study of the postindustrial timescape and its impact on the domestic sphere. *Time and Society* 18 (2–3): 284–305.
Lazzarato, M. 1996. Immaterial labor. In *Radical thought in Italy*, ed. P. Virno and M. Hardt, 133–150. Minneapolis: University of Minnesota Press.
Leonard, S., Fraser, N. (2016). Capitalism's crisis of care. *Dissent*. Retrieved 29 Jan, 2020, from https://www.dissentmagazine.org/article/nancy-fraser-interview-capitalism-crisis-of-care
Lombardozzi, L., and F.H. Pitts. 2019. Social reproduction, social form and social policy: Basic income, basic services, basic infrastructure. *Capital & Class*. https://doi.org/10.1177/0309816819873323.
Lotz, C. 2014. *The capitalist schema: Time, money, and the culture of abstraction*. Lanham: Lexington Books.
Malm, A. 2018. *The progress of this storm*. London: Verso Books.
Marx, K. 1973. *Grundrisse*. London: Penguin.
———. 1976. *Capital*. Vol. I. London: Penguin.
———. 1981. *Capital*. Vol. 3. London: Penguin.
Meszaros, I. 1970. *Marx's theory of alienation*. Brecon: Merlin Press.
Moor, L., and G. Julier. 2009. Introduction: Design and creativity. In *Design and creativity: Policy, management and practice*, ed. G. Julier and L. Moor, 1–22. Oxford: Berg.
Moore, J.W. 2011. Transcending the metabolic rift: A theory of crises in the capitalist world-ecology. *Journal of Peasant Studies* 38 (1): 1–46.
———. 2017. Metabolic rift or metabolic shift? dialectics, nature, and the world-historical method. *Theory and Society* 46 (4): 285–318.
Nitzan, J., and S. Bichler. 2009. *Capital as power: A study of order and creorder*. New York: Routledge.
O'Brien, D., and K. Oakley, eds. 2017. *Cultural policy: Critical concepts in media and cultural studies*. London: Routledge.
Pitts, F.H. 2015a. Creative industries, value theory and Michael Heinrich's New Reading of Marx. *TripleC* 13 (1): 192–222.
———. 2015b. Form-giving fire: Creative industries as Marx's 'work of combustion' and the distinction between productive and unproductive labour. In *Reconsidering value and labour in the digital age*, ed. E. Fisher and C. Fuchs, 246–260. London: Palgrave Macmillan.
———. 2015c. A hidden history: Defining and specifying the role of the creative industries. *Creative Indian Journal* 8 (1): 73–84.
———. 2015d. Creativity and the commodity in the automobile industry. *International Journal Cultural Creative Indian* 2 (2): 32–41.

———. 2016a. Creativity and power in freelance creative work. In *Virtual workers and the global labour market*, ed. K. Randle and J. Webster, 139–159. Basingstoke: Palgrave.

Pitts, F. H. (2016b). *Capital as power in the creative industries: A case study of freelance creative work in the Netherlands*. Working Papers on Capital as Power

———. (2017a). *Marxian value theory and the 'crisis of measurability': A case study of work in the creative industries in the UK and the Netherlands*. Doctoral dissertation, University of Bath

Pitts, F.H. 2017b. Creative labour, before and after 'going freelance': Contextual factors and coalition-building practices. In *The new normal of working lives: Critical studies in contemporary work and employment*, ed. S. Taylor and S. Luckman, 87–107. London: Palgrave Macmillan.

Pitts, F. H. (2019). SMart solutions for the self-employed beyond the 'British Way'. *Migration Mobilities Bristol*. Retrieved 29 Jan, 2020, from https://migration.blogs.bristol.ac.uk/2019/04/29/smart-solutions-for-the-self-employed-beyond-the-british-way/

Pitts, F. H., Jean, E., Clarke, Y. (2019). Sonifying the quantified self: Rhythmanalysis and performance research in and against the reduction of life-time to labour-time. Capital Class. https://doi.org/10.1177/0309816819873370

Postone, M. 1993. *Time, labor and social domination: A reinterpretation of Marx's critical theory*. Cambridge: Cambridge University Press.

Ross, A. 2003. *No-collar: the humane workplace and its hidden costs*. New York: Basic Books.

———. 2008. The new geography of work: Power to the precarious? Theory. *Culture and Society* 25 (7–8): 31–49.

Ross, P., Bird, A., Pitts, F. H., Crowley, L. (2017). *A new way of working for the self-employed – SMart*. coops.consultancy. Retrieved 29 Jan, 2020, from http://www.alex-bird.com/a-new-way-of-working-for-the-self-employed-smart/

Ryan, B. 1992. *Making capital from culture*. Berlin: Walter de Gruyter.

Schmidt, A. 1971. *The concept of nature in Marx*. London: New Left Books.

Sohn-Rethel, A. 1978. *Intellectual and manual labour*. Atlantic Highlands: Humanities Press.

Stahl, M. 2014. Specificity, ambivalence and the commodity form of creative work. In *Theorizing cultural work*, ed. M. Banks, R. Gill, and S. Taylor, 85–98. Abingdon: Routledge.

Williams, R. 1965. *The long revolution*. London: Penguin.

———. 2005. Advertising: The magic system. In *Culture and materialism*, 170–195. London: Verso.

Xhauflair, V., B. Huybrechts, and F. Pichault. 2018. How can new players establish themselves in highly institutionalized labour markets? A Belgian case study in the area of project-based work. *British Journal of Industrial Relations* 56 (2): 370–394.

Yakura, E.K. 2001. Billables: The valorization of time in consulting. *The American Behavioral Scientist* 44 (7): 1076–1095.

Zanoni, P. 2018. Belgium: Reinvigorating the self-regulated labour market model. In *Work in the digital age: Challenges of the fourth industrial revolution*, ed. M. Neufind, J. O'Reilly, and F. Ranft, 317–332. Lanham: Rowman and Littlefield.

———. 2019. Labor market inclusion through predatory capitalism? The 'sharing economy', diversity, and the crisis of social reproduction in the Belgian Coordinated Market Economy. In *Work and labor in the digital age*, ed. S.P. Vallas and A. Kovalainen, 145–164. Bradford: Emerald Publishing.

Chapter 10
'You're Always On, and You're Always Lively': Young People and Creative Work

Anthony Killick and Kate Oakley

10.1 Introduction

The policy rhetoric of the creative industries is Janus-faced. On one hand it is told as a story of high tech, high skill, mobile labour, providing jobs and growth, even in the dismal landscape of the UK economy (DCMS 2018). On the other, it has often been associated with 'regeneration' (LGA 2019), the promise of breathing new life into areas bereft after decades of deindustrialization and further scarred by years of austerity. In the latter case the pitch is about place, not mobility, and about communities and people excluded from the first version of the creative industries. The growth of the creative industries is strongly associated with the expansion of higher education (HE) and, while the 'artistic' professions were traditionally associated with learning outside of formal education, they nowadays feature a highly graduate workforce—over 62% in the UK creative industries (DCMS 2018). Yet the rhetoric continues to associate the creative industries with the process of upward social mobility (Littler 2018), the deeply entrenched idea that 'talent' lies everywhere and the need for cultural expression is widespread and just waiting to be released. Recent years have seen the idea of creative industries fused with digital technologies of one sort or another, particularly by policymakers—to the extent that the UK's onetime Culture Ministry now had 'digital' as its first priority. This has produced a downplaying of the other parts of the creative industries—crafts for example or fine art—in favour of AI, robotics and videogames. On the one hand therefore we have creative hubs, business parks and high tech incubators (HMG 2018) and,

A. Killick
Liverpool John Moores University, Liverpool, UK
e-mail: a.w.killick@ljmu.ac.uk

K. Oakley (✉)
University of Glasgow, Glasgow, UK
e-mail: kate.oakley@glasgow.ac.uk

K. Oakley, M. Banks (eds.), *Cultural Industries and the Environmental Crisis*,
https://doi.org/10.1007/978-3-030-49384-4_10

on the other, we have Cities of Culture, culture-led regeneration and apprenticeships schemes (LGA 2019).

> As the Bazalgette report on the creative industries in 2017 put it,[1] (Bazelgette 2017, p. 3)
> A diverse workforce is pivotal to the innovative and creative processes that give the UK Creative Industries their global competitive advantage. Moreover, under-represented groups constitute an untapped domestic pool of raw talent that, if provided with the relevant skills, could be used to fill the skills gaps facing the Creative Industries.

The language here is deeply problematic, the class and racial overtones of 'raw talent,' are stark, while the implication that such talent can be 'used' to fill skills gaps denies agency or even dignity. But it neatly summarizes the tension at the heart of this policy rhetoric and indeed the whole approach. We know that the benefits of creative industry employment accrue largely to the already well connected and well educated (Friedman and Laurison 2019), but the myth of meritocracy needs to be maintained, in part at least, for policy legitimacy (Littler 2018). The creative industries have many of their roots in the publicly-subsidized arts sectors, or in public service media such as the BBC, and their claims to public legitimacy and support are thus different from those of other sectors of the economy. We may direct subsidy to aerospace or pharmaceuticals, and we are willing to heavily subsidize the financial services industry in times of crisis, but in these cases the expectations are purely for economic growth—social amelioration is not anticipated.

The last 5 years or so have witnessed an explosion of research, both in policy and academic circles, that demonstrates the highly unequal nature of the creative industries with its exclusions of class, gender, age and ethnicity (Friedman and Laurison 2019; O'Brien et al. 2016; Saha 2018). There are also deep regional inequalities. Commercial creative production is highly concentrated. Together with the South East region that surrounds it, London accounts for half of all cultural sector employment in the UK, has 40% of the UK's cultural workers, and a third of all its businesses in that sector (Oakley et al. 2017).

There have been some, so far fairly limited, policy attempts to address these concerns. Limitations on unpaid work, particularly internships, largely drawing on existing minimum wage legislation, have been enacted, as unpaid work clearly acts as barrier to entry into employment (Oakley 2014). A strong rhetoric of 'diversity' accompanies most policy pronouncements in this field—Parliamentary groups and taskforces abound—with as yet limited evidence of success. And growing concerns about student debt, skills shortages in some technical areas and evidence of graduate un- or under-employment means that policymakers are also now looking at non-graduate ways into the creative industries through apprenticeship schemes in particular.

The concern of this chapter is how these conflicts play out in the aspirations and understanding of young people who want to enter the creative industries. For those yet to start work, what once may have seemed like a relatively straightforward transition from some form of formal education into the world of work has become

[1] An independent policy review commissioned by the UK Government in 2017.

considerably more complex (Sefton-Green et al. 2019). The compact between education and the workplace has been shattered by competition, de-industrialization and growing regional inequalities and is further tested by automation and AI (Srnicek and Williams 2016). Short-term jobs, zero-hours contracts, extended internships; all of these markers of precarity are common across the workplace. It is in this—highly uncertain—context that young people are contemplating work in a creative industries sector that has all of these features and more. But it is also a workplace that still comes with considerable positive rhetoric, seeming to offer, if not security, then the possibility of self-expression and fun. There is a growing literature on transitions into cultural work (Allen 2020; Allen and Hollingworth 2013; Campbell 2018; Taylor and Luckman 2020), much of it critical of the nature of the creative industry labour market and its betrayal of the promises of 'creativity.' While not dissenting from this critique, Taylor also asserts the importance of what she calls the 'practitioner concept,' cultural workers' self understanding which she found often differed markedly from that of academics or of policymakers. In analyzing our interviews we have also drawn on this idea.

The research is part of a programme of work we have been carrying out on the role of arts and culture in what this volume refers to as Sustainable Prosperity—an economy that enables us to live well within planetary constraints. As we have argued (Oakley and Ward 2018) the idea of sustainable prosperity looks different in different geographic contexts and serious social inequality is one of the strongest barriers to a more sustainable future. You would not, in many ways, start from here. But here we are and if we want to advocate that arts and culture can play a positive role in such a society, we need to understand both why the creative industries' vision is so antithetical to this, but also, and importantly, why it has such a hold on young people's imaginations. After all, although awareness of the inequalities and damages of these sectors is very widespread, many young people still want to work in these sectors. Creativity, the chance to work in a field that offers self-expression, the challenge of fulfilling work—these are not to be dismissed as human goals, and are indeed something that any model of sustainability needs to take seriously. So we need to find ways to hold on to the promise of creative work while ensuring that not only is it more widely available, but that we also pay attention to conditions of work and ownership of labour within these industries.

The chapter is based on focus groups in very different parts of the UK—from the hyper-gentrified inner city to rural areas and a de-industrialized town. Throughout this project we have conducted research in Islington in inner London, Stoke-on-Trent in the North of England and Hay-on-Wye, a rural town on the English/Welsh border. We also attempted to involve young people from different class backgrounds, with focus groups taking place in multiple locations in both Stoke and Islington. Given the small population in the Hay area, we carried out just one focus group at the Hereford College of Art (HCA) which featured a group from ABC1 backgrounds. The common element for the young people, aside from age, was a stated desire to work in the creative industries.

None of our young people were at University, they were in a mix of school and sixth form colleges or in some cases, out of school and unemployed, and, while some were planning to go to University, others were looking to enter directly by starting their own small businesses or via apprenticeships and other non-graduate pathways. While other work in this area focuses on young people either in specialist education or in the early years of work (Campbell 2013), our focus groups contain a mix of young people in and out of formal education. As we shall see, their knowledge of what constitutes the creative industries or how one might go about trying to get a career in them varied hugely, but they were united in their aspirations to 'be creative'.

10.2 Escaping the Everyday

Scholars such as McRobbie (2016) have written extensively about the enthusiasm that many young people hold for creative work. She has famously described this as 'passionate' work, which, as she notes, lives alongside and sometimes feeds off, the economically precarious nature of such work. As one young man in our Hay focus groups put it, 'it's like an escape route from the mundanity of life, really.' (Male, Hay). Such sentiments—lack of routine, of mundanity, the possibility of change and of individual expression—were common across all the groups, despite the other differences between them. A young man in Stoke says,

> It's always different. You're never doing the same thing twice. With photography you're not going to take the same picture 200 times, or even if you did it'd be different each time. (Male, Stoke)

This possibility of changing work is strongly linked to an idea of personal agency, and being able to change your life: 'With a lot of the CI you can take a different standpoint on life, so you can look at it differently' (Male, Islington).

What was notable about these comments was that only in a minority of cases, was passion expressed for a particular art or cultural form. Some young people did identify with film, fashion, photography or videogames, but for many it was the idea of creativity, the *creative* industries that held appeal. Scholars have spent much time debating the coherences or lack of coherence of the creative industries as an economic construct (Pratt 2008), but it has always been clear that the term has worked rhetorically.

> I think 'career' is very vague in itself, business office all day long, they're sat on their desk doing their work, and it's not very social. It's not very fun. Whereas in all of the arts industries I think it's called a 'creative career' because you're always on, you're always lively. And no matter what you're doing, you're doing something with people, and it's always new and fresh. Every day you'll be doing something completely different. (Male, Hay)

This was frequently contrasted with the '9–5', where 'someone else is telling you what to do': 'I think the difference between this [the CI] and a regular, 9–5 office job is that people who want to go into the CI have passion' (Male, Islington).

These sentiments are widespread across the literature on creative labour (Hesmondhalgh and Baker 2011; McRobbie 2016). What is evident in our research was the degree to which they are articulated by those yet to enter the workplace, and hence how early such notions start and the way that class background sharpens and intensifies these sentiments. As McRobbie has noted in her work on young female creatives (McRobbie 1998), young people often make comparisons between the work that their parents did and their own aspirations. For working-class respondents this was often, and not surprisingly, about doing a more interesting job, with more possibly of self-fulfilment, and while self-fulfilment was also an aspiration for middle class respondents, the need to avoid the fate of previous generations was less keenly expressed. Working-class youngsters, particularly in our Stoke focus groups, who are both socially and geographically remote from concentrations of creative industry work, expressed aspirations that might seem painfully unrealistic.

> When I was at the job centre looking for creative jobs, they told me to focus on 'every day jobs', and also I was working part-time in a job, volunteering at something I enjoy doing, and they were pushing me to try and get a full-time job doing anything. They just want to tick boxes, but they don't really help you. They don't help you look for creative jobs. They don't help you get work experience in creative jobs. (Woman, Stoke)

> I've been in the situation where, like I say I have an A-level in media, I've looked on job-sites, the job centre, and there's not an awful lot in the media job wise out there. For me as an individual now, I would work in any section of the media, creative writing, blogging, anything. I couldn't seem to find anything. (Woman, Stoke)

In such cases, the idea that 'anyone can make it' is powerful and success stories are often touted:

> Individuality is the main thing with me. My mum supports me, but because my mum grew up at a time when, if you were designing your own clothing, you were selling it. With me, I actually want to start my own clothing line and get it big. Like I've said, people in the old days used to sell clothes out of the back of a van. That's how Nike started. They bought a load of trainers with no logo on them, stitched on a tick, sold them out the back of a van, and now he's a multimillionaire. So if he can do that, I can do it. Anyone can do it. (Male, Stoke)

It is possible to read this as a lack of understanding about what the 'creative industries' are and how they work; indeed such sentiments are often expressed in the skills and careers literature, where the need to lay out pathways and clarify entry routes into the sectors is often advanced.[2] There is something in this. It was evident that the working-class youngsters in our groups—those who were furthest from knowing anyone who worked in the creative industries—were more likely to express excitement about the broad idea of these sectors, allied to a lack of knowledge about how they operate. Middle-class respondents have both greater awareness and, in some cases, a more realistic understanding.

> Well for example on films or television production there's many different roles, like editor, runner, cameraman, sound operator, director, producer, editor. It's quite an extensive list … There's a variety of jobs that you could take to get yourself into these industries. (Male, Hay)

[2] See for example https://www.screenskills.com/careers/job-profiles/.

> At the college [HCA] they're very realistic about things. So, they're not just going to say 'you're going to make it.' It's like 'well, two or three of you might,' and that's the nature of the industry. (Male, Hay)

But to read this *just* as lack of industry nous, to be remedied by better careers advice in schools, is to miss the powerful cocktail of discourses that lie behind such expressions. As Allen and Hollingworth (2013) have argued, these discourses are not only embedded in government policy, but amplified by a media fixated on celebrity and success.

One young woman reflected this argument

> I don't feel like anyone in my life has really done anything to the extent that it's, like, 'shit, they've really done something.' I inspire myself … When I see a celebrity I feel like you do see all the stuff they have, it's different. Inspiration is different to just seeing something and thinking 'oh they have a really nice car and they're so talented.' (Female, Stoke).

Mindful of the practitioner concept, it is of course important to try and understand what is going on here. Despite a decade of austerity, many more years of increasingly precarious labour markets and what some researchers have argued is the internalization of that precarity (Berry and McDaniel 2020), young people still refer to the '9–5' or the 'boring office job,' as something from which to escape. For some of these young people, even their parents will not have experienced the mundane workplace or the office job—insecurity, not stasis is generally viewed as problem of the current labour market. Given their ages, it may well be that it is the experience of schooling that is what they are reacting against. But it is important not to lose sight of this desire for change or to see it simply as a misguided reading of career prospects.

10.3 Working Hard and Being Judged

Alongside the rhetoric of creativity and escape from the everyday, there was the realization for most young people that these 'desirable' industries were difficult to get into and required commitment. And alongside stories of instant fame and wealth, there was also acknowledgement of financial constraints and—in a few cases—the way this interacts with class and background. As one young man in Islington put it,

> When you're an artist you'll always have two jobs, because first you need to support your art, you need to buy the equipment, which is expensive. So you're always going to have to do another job to get money to support yourself. (Male, Islington)

A working-class woman, also in Islington, noted,

> It's all good and well if you have enough money to be able to produce what you need to produce and then put it out there. But if you don't have that money it's a problem. I don't think there's a lot of scholarships or bursaries given to students in the CI, not in fashion anyway. (Female, Islington)

The class disparities in the creative industries have been the subject of burgeoning academic research in recent years (Friedman and Laurison 2019) as well as increased media coverage, and some of this debate was echoed by our respondents.

> [It's not just about] how people judge your artwork, but how they judge you. They might not like you, they might just look down at you … The way you present yourself, if I was to go into a meeting in a tracksuit and high tops they won't even listen to me speak. They'll just judge me straightaway. I could speak so eloquently but they would just look at what I wearing and say 'oh, he's one of them people' … that's just the way it is … it's a fine line to cross, staying true to yourself in an environment where people are so judgemental. (Male, Islington)

The adage that 'it's not what you know but who you know', was expressed by more than one respondent, but these ideas of class-based judgement or the relevance of social contacts was rarely seen as an issue of justice or injustice—more of a recognition of 'how things are'. Respondents prized themselves on their realism in recognizing that acquiring qualifications and skills would not be enough—that they would need something 'extra'. For the Hay-on-Wye group, a largely middle-class group at a specialist sixth form college, much of this was reflected in what their tutors told them. Alongside specific skills training, much emphasis was on 'hints and tips', ways to demonstrate the much desired 'positive' outlook alongside the coded cultural skills of the creative sectors (Friedman and Laurison 2019). As Banks argues (Banks 2017, p. 84), these are the 'social appearances and dispositions that would invite selection as talent.' As he points out, despite the claims of meritocracy advanced for the creative industries, the lack of clear, formal routes in often means that aesthetic judgements ('whose work do you admire?)', certain levels of class-based self confidence and homophilic prejudices determine both admittance to and advancement within, the creative sectors.

> We had an artistic director come in and speak to us and she said to us 'a lot more comes into it than you think. If you're the first one in the audition room at 9am you haven't got the part', because they [casting directors] haven't had anything to eat, they haven't had their coffee. It's ridiculous. She said to us 'if you have an audition at 11 and you look right you've got the part, because they've had some food, they've woken up a bit, and they've seen a bunch of people already' … whoever goes in at the right time gets the part. (Male, Hay)

> Name dropping, going back to this thing that you've always got to network, to name drop all the time will make people remember who you are … not making yourself look like a pretentious idiot, but that's the problem. You don't want to just look like you're showing off and people won't like you, but you've got to show off. So social skills are massive as well. (Male, Hay)

While all groups recognized that the entrance to such industries was often non standard and the links between formal educational qualifications was less clear than in other sectors, understanding of how this actually operates were variable. There was much debate about the relevance of qualifications—some students were doing A levels with University in mind and, others were doing BTEC, while others were working on their own projects in film, photography or fashion. The importance of 'networking' was acknowledged by all, but understanding of how that actually operates and access to such networks was hugely disparate.

A small number of our middle-class interviewees had parents or relatives who worked in the creative industries or around the margins of them. For them the possibility of a personal 'in' was highly valued, but even where this was not a possibility, there was a high level of awareness of the need for face-to-face networking, for relevant work experience and for social contacts.

> It's not necessarily what you know, it's who you know, in that there's never going to be a time in an amazing director or producer's career where they haven't had to brown nose someone for years and years just to get to where they want to be. (Male, Hay)

> As an actor, as with anyone in the performing arts you have to sell yourself. So online if anyone can search for you in an instant that's got to be your best self, because if it's not ... if there's something on there that a casting agency doesn't want to see, then they're not going to hire you ... There's a difference between online and in person [networking] because online is just static, you can put something there and just leave it, whereas person to person you have to keep going on with it. (Male, Hay)

These young people's understandings correspond much more closely to the notion of a 'scene' or milieu as described in work on music and indie games development. Writing about Austin in Texas, Watkins and Lombana-Bermudez (2019, p. 30) describe the mix of what they call ' techno capital'—familiarity with digital means of production and with platforms—with social capital—contacts, relationships and networks all taking pace within physical spaces that allow young people, 'to make real-world connections, collaborate, share their work, and build community.' As they argue, it is this combination that allows scenes to develop and thrive and while such scenes are marked by exclusions and exploitations, they are also capable of offering a sense of community, however fragile. Even in high tech fields such as indie game development, this was the case. As they argue, young people's cultural participation is, contrary to popular opinion, very rarely digital-only. What was striking in our interviews was that, while this was true for many of our interviewees, who recognized the need for face-to-face interactions and cultural consumption, for the most economically-deprived of our young people the promise of 'creative' has become the promise of 'digital'.

For those young people, 'working in the creative industries' had taken on an inflection removed from place, from personal contacts and sometimes from personal creative enthusiasms. This was particularly true for some of our Stoke focus groups, which contained some of the poorest young people we spoke to. We asked where people thought they may have to move to in order to realize their ambition of working in the creative industries. Young people in Islington, from a variety of social backgrounds were, perhaps not surprisingly, London-centric. They were aware that London dominated the UK in terms of creative industry employment and, while some talked about working abroad (for example, New York), none mentioned going elsewhere in the UK. The benefit of physical proximity was something of which they were keenly aware. Our Hay group could be seen as the most geographically isolated; Stoke is, after all, less than 2 h from London by train. But they were well aware and largely sanguine that they would have to move to London or another large city, and for the most part that transition could be eased by family members or other family connections already located there. For the young people in Stoke, questions about possible work locations were met by a variety of answers—Italy ('its got old buildings'), Dubai or Mexico—though more from enthusiasm than any sort of realistic appraisal. With little immediate prospect of leaving Stoke, and little idea of how they would manage the transition if they did, they separated themselves in other ways, imagining themselves to be working in the creative industries, whereas others around them are 'boring,' 'basic' and so on. Geographically remote from the centres of cultural employment and socially isolated, in part by poverty, the idea of creativity had become—in all senses—virtual. Working in the creative industries has become working in social media.

10.4 Social Media Celebrity

Creative industries policy has often stressed the importance of place, partly as discussed earlier, because of its appeal to the regeneration and 'turnaround' of depressed regional economies. Other chapters in this book by Paula Serafini and Cecilia Dinardi express the global nature of that appeal. Geographers such as Pratt (2002) have argued compellingly that, *contra* to the pronouncements about the 'death of distance' that digital technology would bring about (Cairncross 1997), the nature of creative work—with its premium on tacit knowledge and social relationships, the importance of face-to-face and 'thick' labour markets—is such that close physical contacts would remain important in creative production and hence places that have economic advantage in such industries would retain them. The rise of social media however, the growth of influencers and Youtubers, the dominance of streaming platforms in music and film (Hesmondhalgh 2019), has given the 'death of distance' notion a second life however (Ashton and Patel 2018) and, in the imaginaries of some young people, working in the creative industries has moved online.

> Right now the internet gives people a platform where they can put up their work and get people to see it. So now you have this platform where you can get people to see what you're capable of producing. Back in the day, before the internet, that was very difficult. Like with music, you had to get on the radio. But now, with stuff like Spotify and Soundcloud you can just upload your stuff on there and grow a fanbase. (Male, Islington)

> I put films on YouTube because it's a free platform and it gets you exposure anywhere. If I had the will to, I could email the head of the biggest film company in the world. I don't need to be outside of his door physically. (Male, Stoke)

While such aspirations are not unusual, indeed engaging with social media has increasingly become part of cultural workers' practices (Duffy 2016), what was apparent was that the further young people were from any actual contact with the creative industries—either through family connections or specialist education—the more that digital platforms loomed large in their minds as the solution. This was not *just* about geographical remoteness; it was expressed by young people in Islington, as it was in Stoke, but class and wealth play a role. In addition to seeing digital platforms as a way to get their work shown, for a minority simply being online and hopefully getting famous online—getting 'blue-ticked', becoming an influencer—was the goal. Faith in the promise of online celebrity was much stronger among those for whom it may be the only way in. Young people, faced with incredibly limited employment opportunities are having to make their own, and the world of the online celebrity seems appealing and almost reachable. It offers to free them not only from place and the need to make contacts or get qualifications, but in some cases from the need to think about cultural production at all,

> I actually use instagram as a business page. I actually watch what goes on my instagram, how many people view it, each post, everything, which I find quite interesting actually. It's mainly [my] daily life. It's almost like a bit of a blog, all the stories that are on there. I have quite a busy lifestyle anyway. Like, I am one of those that goes out to work every day. I enjoy going to the gym. So it's anything and everything, for people to see the reality of my life, and then probably relate to some of the stuff that I'm doing … It's just a public figure that someone can relate to and follow. (Female, Stoke)

> People draw pictures of celebrities and put them up on Instagram, and then the celebrities screenshot it and then they pay them. (Female, Islington)

> People who go through YouTube, they've got products out there now. Zoella, on YouTube, she's got bath and home. There's someone out there who's got hairsprays and stuff. At first it just starts as merch, t-shirts and stuff like that. Then they go on. They've had books out, and everything else like that. So it's not just that it gets followed through you making your videos. You can go onto writing your books and everything else that you want to do. (Female, Stoke)

As discussed, added to the promise of creativity that has been at the forefront for decades now, is the promise of digital also celebrated by policymakers, educators and thousands of glossy brochures. But just as it has become easier than ever before to make and distribute cultural content, it may never have been harder to sustain a living as a cultural producer. Digital tools and distribution platforms, social media and apps means that many more of us can create and share content of all sorts, but these same technologies that have ripped up business models, have also ripped up livelihoods. New business models arise of course and industries re-form—a music industry based on a mix of live performances, merchandizing and music streaming is forming (Watkins and Lombana-Bermudez 2019). But the indications are that these new industries are based on the same mix of face-to-face and remote connection, of skills, confidence and hustling, of social contacts and social media. The digital dream that is being sold alongside the creative dream is just, if not more, likely to disappoint, and it is as likely to break along the lines of gender, race and class.

10.5 Conclusions

The views of young people about work in the creative industries may seem a long way from the overall topic of this book. The environmental crises, climate change, what society will be like in a climate-chaotic world; these were not the subjects under discussion. But the CUSP project, of which it is a part, is concerned with how we might live and work well, if not better, under conditions of extreme resource constraint. And reimagining both work and creative production are at the heart of that project.

John Bellamy Foster (2017) in an essay on good work and sustainable prosperity draws on a long tradition of writing about creative work as a form of unalienated labour, a way to reclaim work both from the idealists of the permanent leisure society (Gorz 1985; Bregman 2017) and from the degradation of work under contemporary capitalism. While we agree that a society without work is not necessarily a desirable one, his faith in creative work as good work is one that does not always survive much examination. And yet it persists. The thread that links Bellamy Foster's thoughtful defence of work to the latest outputs from creative industry policymakers is a long one, but it draws on some of the same idea about creative work, self-expression and the fulfilment of human potential.

It is also an idea that is reflected in the views of our interviewees. It is tempting to dismiss this as what Morgan and Nelligan (2017) call the creativity 'hoax,' a false promise of late capitalism that will for the most part lead to disappointment. But, as we have noted earlier, it is also important to take these views seriously—and to try and understand what is being appealed to here. As Taylor has found when interviewing established cultural workers over many years, this emphasis on transcendence and an escape from the mundanities of life is common and is often stressed more than conventional success markers such as fame or money (Taylor and Littleton 2012; Taylor and Paludan 2019). This persists throughout careers which are often far from uniformly successful, as does the notion of a 'creative self', separate from the particular practice that one is following. Taylor argues that this creative self can endure throughout setbacks and precarious employment and that even during the 'drudge' work that people often undertake to support their creative practice, 'the creative self is not being mobilised, but neither is it erased or invalidated' (Taylor and Luckman 2020).

The young people in our focus groups came from different social, economic and geographic backgrounds in a world where such differences are rapidly becoming chasms. The appeal of creative work, the exhortation to 'be creative' is common across all of them, as is the recognition that this may carry with it risk, insecurity and the need to 'hustle'. Clearly the ability of young people to undertake this sort of work—the degree of economic capital behind them, the right sort of networks and connections, even an understanding of what these industries are and how they work—is very uneven. It would be reasonable to conclude that what we need is better access to the creative industries: more apprenticeships, more 'diversity' schemes, online learning, shorter degree courses and so on. Some of this is undoubtedly desirable, but what is also apparent is the need to combine this with re-thinking the nature of work itself, not just work in the creative industries.

In part this is about reviving the different traditions of cultural work—of guilds and trade unions, of artists' co-ops, collective ownership and communal production—that have been a part of it alongside the individual genius or the atomized precarious worker. A focus on these traditions of work, on working conditions and crucially on ownership and control, is needed before creative work can be part of a sustainable prosperity. But it may also require attention to the ideas of transcendence and the creative self and at what possibilities lie within that. Commentators are divided about the future of work—and while only the most optimistic foresee a utopia of untrammelled leisure time (Srnicek and Williams 2016), the majority see either under or unemployment and worsening inequalities (Spencer 2018). It can thus be difficult to hang on to an optimistic reading of creative work as unalienated.

References

Allen, K. 2020. Young women's aspirations and transitions into, through and away from contemporary creative work. In *Pathways into creative working lives*, ed. S. Tayor and S. Luckman, 50–61. London: Palgrave MacMillan.

Allen, K., and S. Hollingworth. 2013. 'Sticky subjects' or 'cosmopolitan creatives'? Social class, place and urban young people's aspirations for work in the knowledge economy. *Urban Studies* 50 (3): 499–517.

Ashton, D., and K. Patel. 2018. Vlogging careers: Everyday expertise, collaboration and authenticity. In *The new normal of working lives*, ed. S. Taylor and S. Luckman, 147–170. Cham: Palgrave Macmillan.

Banks, M. 2017. *Creative justice. Cultural industries, work and inequality*. Lanham: Rowman & Littlefield.

Bazelgette, P. (2017). *Independent review of the creative industries*. Retrieved 4 Feb, 2020, from https://www.gov.uk/government/publications/independent-review-of-the-creative-industries

Bellamy Foster, J. (2017). *The meaning of work in a sustainable society: A Marxian view. CUSP essay series on the morality of sustainable prosperity*. Retrieved 4 Feb, 2020, from https://www.cusp.ac.uk/themes/m/m1-3/#1475182667098-0328ae0f-4bcbf2c7-159eee31-91cf

Berry, C., McDaniel, S. (2020). Post-crisis precarity: Understanding attitudes to work and industrial relations among young people in the UK. *Economic & Industrial Democracy*. https://doi.org/10.1177/0143831x19894380

Bregman, R. 2017. *Utopia for realists*. London: Bloomsbury Publishing.

Cairncross, F. 1997. *The death of distance. How the communications revolution will change our lives*. London: Orion Business Books.

Campbell, M. 2013. *Out of the basement: Youth cultural production in practice and in policy*. Montreal: McGill-Queen's University Press.

Campbell M (2018) 'Shit is hard, yo': Young people making a living in the creative industries. Int J Cultural Policy. https://doi.org/10.1080/10286632.2018.1547380

DCMS. (2018). *Economic estimates*. Accessed 5 August, 2019, from https://assets.publishing.service.gov.uk/government/uploads/system/uploads/attachment_data/file/759707/DCMS_Sectors_Economic_Estimates_2017__provisional__GVA.pdf

Duffy, E.B. 2016. The romance of work: Gender and aspirational labour in the digital culture industries. *International Journal of Cultural Studies* 19 (1): 441–457.

Friedman, S., and D. Laurison. 2019. *The class ceiling. Why it pays to be privileged*. Bristol: Policy Press.

Gorz, A. 1985. *Paths to paradise: On the liberation from work*. London: Pluto Press.

Hesmondhalgh, D. 2019. Have digital communication technologies democratized the media industries? In *Media and society*, ed. J. Curran and D. Hesmondhalgh, 101–120. London: Bloomsbury Academic.

Hesmondhalgh, D., and S. Baker. 2011. *Creative labour, media work in three cultural industries*. London: Routledge.

HMG. (2018). *Industrial strategy. Creative industries sector deal*. Retrieved 4 Feb, 2020, from https://www.gov.uk/government/publications/creative-industries-sector-deal

LGA. (2019). *Culture-led regeneration. Achieving inclusive and sustainable growth*. Local Government Association

Littler, J. 2018. *Against meritocracy: Culture, power and myths of mobility*. London: Routledge.

McRobbie, A. 1998. *British fashion design: Rag trade or image industry?* London: Routledge.

———. 2016. *Be creative. Making a living in the new culture industries*. Cambridge: Policy Press.

Morgan, G., and P. Nelligan. 2017. *The creativity hoax. Precarious work and the gig economy*. London: Anthem Press.

O'Brien, D., D. Laurison, A. Miles, and S. Friedman. 2016. Are the creative industries meritocratic? An analysis of the 2014 British Labour Force Survey. *Cultural Trends* 25 (2): 116–131.

Oakley, K. 2014. Good work? Rethinking cultural entrepreneurship. In *The handbook of management and creativity*, ed. C. Bilton and S. Cummings, 145–159. Cheltenham: Edward Elgar.

Oakley, K., and J. Ward. 2018. The art of the good life: Culture and sustainable prosperity. *Cultural Trends* 27 (1): 4–17.

Oakley, K., D. Laurison, D. O'Brien, and S. Friedman. 2017. Cultural capital: Arts graduates, spatial inequality, and London's impact on cultural labour markets. *The American Behavioral Scientist* 61 (12): 1510–1531.

Pratt, A.C. 2002. Hot jobs in cool places: The material cultures of new media product spaces; the case of the South of the Market, San Francisco. *Information, Communication & Society* 5 (1): 27–50.

———. 2008. Cultural commodity chains, cultural clusters, or cultural production chains? *Growth Change* 39 (1): 95–103.

Saha, A. 2018. *Race and the cultural industries*. Cambridge: Polity Press.

Sefton-Green, J., S. Watkins, and B. Kirshner. 2019. *Young people's transitions into creative work*. Abingdon: Routledge.

Spencer, D.A. 2018. Fear and hope in an age of mass automation: Debating the future of work. *New Technology, Work and Employment* 33 (1): 1–12.

Srnicek, N., and A. Williams. 2016. *Inventing the future: Postcapitalism and a world without work*. London: Verso.

Taylor, S., and K. Littleton. 2012. *Contemporary identities of creativity and creative work*. Farnham: Ashgate.

Taylor, S., and S. Luckman. 2020. *Pathways into creative working lives*. London: Palgrave MacMillan.

Taylor, S., Paludan, M. (2019). Transcending utility? The gendered conflicts of a contemporary creative identification. *Feminism & Psychology*. https://doi.org/10.1177/0959353519864390

Watkins, S.C., and A. Lombana-Bermudez. 2019. Being Indie: The DIY ethos and Indie game development. In *Young people's transitions into creative work*, ed. J. Sefton-Green, S. Watkins, and B. Kirshner, 23–31. Abingdon: Routledge.

Chapter 11
The Green New Deal and Cultural Policy

Jonathan Gross and Nick Wilson

11.1 Introduction

In response to the climate emergency a growing number of researchers, journalists, economists, politicians and activists are calling for a Green New Deal (GND). This phrase, of course, deliberately echoes Roosevelt's New Deal of the 1930s, in which the US administration recognized that a rupture on the scale of the Great Depression required co-ordinated governmental intervention in the socio-economic life of the nation. What supporters of a GND advocate is sustained intervention on this scale (and more), with the aim of making the 'greening' of the economy foundational to all aspects of political decision-making.

The idea of the GND was first articulated shortly after the financial crash of 2008, when a group of economists and academics began meeting informally to consider policy responses to that seismic event, and its relationship to the politics of climate change (Pettifor 2019). Over the next decade, these ideas have gained traction to the extent that major political parties are now beginning to place a GND at the centre of their election manifestos (for example, the UK Labour Party 2019). However, notwithstanding the visibility these ideas have gained, the achievement of a GND as official public policy faces enormous political obstacles, not least in those countries in which right-wing populism is in the ascendancy. Perhaps most obviously and damagingly of all, Donald Trump's withdrawal of the USA from the Paris Agreement (United Nations 2015) makes plain the scale of the political challenge.

In the estimation of some commentators, meeting the challenges of climate change requires 'a transformation of our global economy, our society, and our very way of life without precedent in human history' (Rifkin 2019, p. 2; see also Hallam 2020). In this context, what serious role, if any, can be played by cultural policy? The answer might appear to be 'limited, at best.' The green credentials of current

J. Gross (✉) · N. Wilson
Department of Culture, Media & Creative Industries, King's College London, London, UK
e-mail: jonathan.gross@kcl.ac.uk; nick.wilson@kcl.ac.uk

K. Oakley, M. Banks (eds.), *Cultural Industries and the Environmental Crisis*,
https://doi.org/10.1007/978-3-030-49384-4_11

cultural policy are questionable. This includes the failure to address the cultural economy's large carbon footprint (Maxwell and Miller 2017), and prevailing discourses of 'creative economy' which—in their continuing commitment to an underlying model of unsustainable economic growth—not only limit 'cultural imagination' but actively *de-future* (Banks 2018). However, if our social-economic arrangements are to be radically re-futured, as they need to be, that domain of human activity where we exercise our capacities for developing narratives, imagination and new systems for recognizing value (Wilson 2020)—the cultural—will play an essential role.

The climate emergency and the 2008 financial crisis constitute irreparable breaks in prevailing political narratives, and to meet these unprecedented challenges, *new socio-economic stories* must be told (Monbiot 2017; Gross 2019). It is in this context that the starting point for our chapter is to ask, what would it mean for cultural policy to take seriously its potential contributions to addressing 'the most profound set of issues humanity has ever faced' (Morgan and Fullbrook 2019, p. 3)? To offer an initial answer, we propose the need for cultural policy to take up the challenge of the GND. In what follows, we therefore introduce key components of the GND, discuss the nature and distinctiveness of *cultural infrastructure*, and conclude by suggesting what implications this has for telling a new story of cultural policy: one that is committed to re-futuring not only the creative economy, but the economy as a whole.

11.2 The Green New Deal

Conversations about a GND have been taking place since 2008 (Pettifor 2019). From that point, political parties and intergovernmental bodies—from Germany to South Korea to the United Nations Development Programme—began issuing manifestos and official commitments adopting its language (Rifkin 2019). However, it has been much more recently that the idea of a GND has achieved breakthrough into mainstream political discussion. Echoing the vocabulary of her earlier book *No is Not Enough* (Klein 2017), one of the most prominent champions of radical climate action, Naomi Klein, says that it was not until 2018, with the rising visibility of the idea of a GND, that 'there was finally a big bold "yes" to pair with the climate movement's many "no's", a story of what the world could look like after we embraced deep transformation, and a plan for how to get there' (Klein 2019, p.27).

As the activist slogan puts it, the GND is a call for 'System Change, Not Climate Change' (echoed by Pettifor 2019). This means not just nudges on market behaviour via carbon taxes and other regulatory mechanisms, it is 'a major operating system upgrade' (Klein 2019, p. 32) for the economies of the world. Markets will play a part in this, but only a supporting one; and Aronoff et al. (2019) accuse those who use the language of the GND without seeking to transform prevailing socio-economic systems as offering a 'faux Green New Deal', not the real thing. For Klein, one of the conclusions the 2018 report of the Intergovernmental Panel on

Climate Change (IPCC) made unambiguously clear was that humanity will not meet the challenges it faces 'unless we are willing to embrace systemic economic and social change' (Klein 2019, p. 33). For many proponents of a GND, climate action and addressing systemic economic injustice are inseparable. Klein succinctly summarizes the interconnection between the climate and socio-economic crises:

> The idea is a simple one: in the process of transforming the infrastructure of our societies at the speed and scale that scientists have called for, humanity has a once-in-a-century chance to fix an economic model that is failing the majority of people on multiple fronts. Because the factors that are destroying our planet are also destroying people's quality of life in many other ways, from wage stagnation to gaping inequalities to crumbling services to breakdown of any semblance of social cohesion. Challenging these underlying forces is an opportunity to solve several interlocking crises at once. (Ibid., p. 26)

The value of the comparison to Roosevelt's New Deal of the 1930s is that 'it showed how radically both a society's infrastructure and its governing values can be altered in the span of one decade' (Klein 2019, p. 34). Wartime mobilizations provide an alternative precedent. But whilst collective efforts such as World War II and post-war reconstruction demonstrate the kind of ambition and scale that supporters of a GND call for, such previous projects fall short in being a model for what is required now: in part, because they were 'highly centralized top-down transformations' (Ibid., p. 35).

The comparison to the 1930s New Deal is also problematic due to the racism and sexism that characterized its work and housing programmes (Aronoff et al. 2019; Klein 2019; Rifkin 2019). For many of those who back the GND—such as Alexandria Ocasio-Cortez, one of the chief advocates of the GND in the USA—a twenty-first century New Deal must make explicit provision for addressing the injustices of racialized capitalism. For Klein, however, the biggest limitation of the analogy with all previous examples including FDR's New Deal, World War II and the Marshall Plan is that, 'together, they succeeded in kick-starting and massively expanding the high carbon lifestyle of suburban sprawl and disposable consumption that is at the heart of today's climate crisis' (Klein 2019, p. 37). As the IPPC report states, 'there is no historical precedent for the scale of the necessary transitions, in particular in a socially and economically sustainable way' (quoted in Klein 2019, p. 37). With this being the case, whilst the historical models are powerful and necessary tools to explain and champion the magnitude of intervention that is needed, the GND also requires a significant leap of political imagination. The conditions from which such a project is seeking to emerge include the effects of many years of neo-liberalism degrading public services and infrastructures, and denigrating ideas of the public realm and public goods. Notwithstanding their limitations, historical precedents from the first half of the twentieth century do nonetheless 'remind us that another approach to profound crisis was always possible and still is today' (Ibid., p. 38).

At the centre of a GND will be the transformation of 'infrastructure', and it is precisely in the differences between the infrastructure projects of the 1930s and those of the GND that the limits of the comparison with FDR's programme become most visible and consequential. As Jeremy Rifkin (2019) explains, the infrastructure of the Third Industrial Revolution will be highly networked and distributed in nature, and institutions of democratic control and accountability will be essential. The achievement of new infrastructures of these kinds will be a deeply political project. Whilst all of humanity will 'win', collectively, there will be losers. In particular, the fossil fuel companies and their shareholders. It is partly for this reason that Klein argues that 'any administration attempting to implement a Green New Deal will need powerful social movements both backing them up and pushing them to do more' (Klein 2019, p. 261). She goes so far as to suggest that the actions taken by social movements will be 'the single largest determining factor in whether a Green New Deal mobilization pulls us back from the climate cliff' (Ibid.). These movements are about 'building political power—enough to change the calculus of what is possible' (Ibid., p. 262). Here, again, the New Deal of the 1930s proves instructive. Given the civil unrest taking place, the New Deal 'appeared at the time to be the only way to hold back a full-scale revolution' (Ibid.). It was 'pressure from the left, in the form of militant movements and political parties', she writes, that 'delivered the most progressive elements of the New Deal and the Marshall Plan' (Ibid., p. 263).

Many of today's climate activists adopt a revolutionary position that is framed in terms of multiple, simultaneous processes of systemic change. One of the questions that needs to be considered is, what are the systems that matter? Advocates of the GND have argued that the challenges of climate change require a profound transformation of infrastructure: our energy systems, and the modes of communication, transport, work and decision-making with which they are symbiotic. But such systems are 'so basic to the organization of economic and social life', with such far-reaching consequences for how we understand collective human existence, that 'the Green New Deal infrastructure is as much about a change of consciousness as it is about a change in infrastructure' itself (Rifkin 2019, p. 99). What, if anything, does cultural policy have to say (explicitly or implicitly) to these plans for infrastructural transformation? To suggest an answer to this question, let's first consider the rising use of ecological language within cultural policy, cultural policy studies, and the areas of activity in which they take an interest.

11.3 Cultural Infrastructure

Ecological language has, in recent years, increasingly been applied to the arts and to the cultural and creative industries (Barker 2019; Holden 2015, 2016; Howkins 2010; Markusen et al. 2011; Nicholson et al. 2018). As Barker discusses, there are a variety of reasons why writers across a number of fields have made use of this vocabulary, including the need for a language with which to more effectively represent how value is generated, beyond the linearity of the 'value chain'. Across the range of policy and academic texts employing ecological terms, there is plenty of 'terminological elasticity' (Barker 2019), and there remain many conceptual issues to clarify and develop.

Our own work has explicitly linked the use of ecological language to the possibilities for reviving and developing the project of 'cultural democracy' (Jeffers and Moriati 2017; Kelly 1984). In doing so, our research is committed to causal explanation and adopts a realist ontology. We argue that ecological language is needed in order to do justice to how culture, as our system(s) of value recognition (Wilson 2020), is made and re-made: namely, via the interconnections and interdependencies between cultural resources of many kinds (Gross and Wilson 2018, 2019; Wilson et al. 2017; Wilson and Gross 2017). Unlike those discussions of cultural democracy which tend towards characterizing it as a set of fairly specific practices, we suggest that cultural democracy needs to be understood as 'a normative commitment, a system of cultural governance, and a policy programme' (Gross 2019, p. 12), each directed towards the expansion of cultural capability: the substantive freedom to give form and value to our experiences of being-in-relation (Gross and Wilson 2018, p. 5). Here the ecological and the democratic are inseparable. It is only with an expanded sense of cultural resources—and, in particular, their interdependencies and interconnections—that we can appreciate what are the cultural resources that matter, and how they make a difference to people's real cultural opportunities (Gross and Wilson 2018).

In making the case for policy makers to pay attention to more local scales of activity and assets than those typically framed within the 'creative city' script, Bell and Orozco (2020, p. 10) refer to 'the micro-geographies of the arts ecology'. However, they also speak of 'the relational geography of arts *infrastructure*' (Ibid., p. 7, emphasis added) and "cultural *infrastructure*" (Ibid., p. 5, emphasis added), without fully elaborating the intended scope of these infrastructural terms. Some cultural policy makers and practitioners we have worked with directly contrast the language of ecology and infrastructure, expressing a preference for one over the other (Gross and Wilson 2019). But whilst we have made extensive use of ecological terminology, here we want to explain the place of infrastructural language within our overall account of cultural eco-systems, and argue for the importance of *cultural infrastructure* in a specific sense. This, we suggest, has significant implications for understanding the potential role of cultural policy within the overall project of a GND.

The need to better understand the nature and potential of cultural infrastructure can be illustrated via our work with Arts Council England's Creative People and Places (CPP) programme. CPP is an experiment in devolved arts funding. In more than twenty-one locations around England, money is spent by and for people in that area. Working across CPP locations through a combination of interviews, focus groups, workshops and surveys, we compiled an inventory of the heterogenous components of these cultural eco-systems. We identified 54 components in total, including everything from National Portfolio Organisations, to geographic features of the landscape, to car parks, to informal networks, to features of local history. Alongside this range of tangible and intangible 'cultural resources', we identified a series of systemic features of the eco-systems that mediate the relationships between these resources, including the role of the local authority, the presence or absence of institutions of higher education, and the demographic profile of the area (Gross and

Wilson 2019). Drawing on these empirical findings, cultural infrastructure, as we are characterizing it here, refers to all those institutions and resources – including 'relational goods' (Donati and Archer 2015) such as trust – that contribute to holding open spaces for experiencing our relationality: our being-in-relation with the world, its possibilities and value. CPP has been the subject of scepticism from commentators who question the extent to which it has really introduced a step-change in cultural governance (Jancovich 2019). But regardless of the radicalism of these initiatives in their current form, constituting just one experiment in democratic approaches to cultural governance, they help us to make visible many of the individual and systemic elements that make a difference to cultural opportunity.

Here it is useful to introduce the notion of *social infrastructure*, 'the physical places and organizations that shape the way people interact' (Klinenberg 2018, p. 5). As outlined below, we suggest that cultural infrastructure is a sub-set of social infrastructure. Having initially developed his thinking in relation to comparative mortality rates in Chicago neighbourhoods during a heatwave, American sociologist Eric Klineneberg explains that:

> When social infrastructure is robust, it fosters contact, mutual support, and collaboration among friends and neighbours; when degraded, it inhibits social activity, leaving families and individuals to fend for themselves. Social infrastructure is crucially important, because local, face-to-face interactions—at the school, the playground, and the corner diner—are the building blocks of all public life. People forge bonds in places that have healthy social infrastructures—not because they set out to build community, but because when people engage in sustained, recurrent interaction, particularly while doing things they enjoy, relationships grow. (Ibid.)

But not all social infrastructure functions in the same way, and it can be used for a range of social and political aims. In their variations they 'play different roles in the local environment, and support different kinds of social ties' (Ibid., p. 17). For example, those social infrastructures that are specifically designed to promote efficiency 'tend to discourage interaction' (Ibid., p. 18). Klinenberg also suggests that, 'given the world's cultural diversity, it's not a surprise that there is great variety in the kinds of social infrastructure that people find essential' (Ibid., p. 19). Similarly, the characteristics of social infrastructure develop over time. Infrastructures symbolize epochs, reflecting 'dominant ideas about how to organise economy and society' (Ibid., p. 232).

The word infrastructure was first used in English in the late nineteenth century and accelerated in usage in the post-World War II era. It is 'a collective term for the subordinate parts of an undertaking, substructure, foundation' (*Oxford English Dictionary*, quoted in Klinenberg 2018, p. 14). Part of the character of infrastructure, Klinenberg comments, is that it normally goes unnoticed, becoming most visible precisely at the moment it starts to fail. He argues that the importance of the physical and social environment has been grievously under-recognized. Social infrastructure requires investment, and when it degrades, 'the consequences are unmistakable. People reduce the time they spend in public settings and hunker down in their safe houses. Social networks weaken. […] Distrust rises and civic participation wanes' (Ibid., p. 21). He argues that building new social infrastructure is 'just as urgent' as addressing levees, bridges and railways (Ibid., p. 24). With the crumbling of many infrastructural systems, and with the effects of climate change increasingly visible, 'vulnerability is in the air' (Ibid., p. 10). This situation, he suggests, is not sustainable.

For Klinenberg, the importance of this issue can hardly be overstated. 'The future of our democracy is at stake' (Ibid., p. 176). Spending time in places like libraries involves living with difference, and constitutes a kind of 'democratic experiment' in itself (Ibid, p. 46). This is quite unlike spending time online, whatever arguments might be made for the possibilities of the digital public sphere. Klinenberg's ethnographic work in New York's public libraries show these to be spaces characterized by practices of care, non-transactional relationships, the recognition of dignity, and the assumption of human potential. However, the design of social infrastructures can be used to enforce unequal power relations. Oppressed groups therefore require spaces in which 'counterpublics' can form. These are 'essential tools for civic engagement in unequal societies, because they give marginalized groups the private forum they need before engaging other groups' (Ibid., p. 160). Many varieties of social infrastructure are needed, including, in particular, that sub set of social infrastructures that we are naming here as cultural infrastructures. Those, specifically, that support cultural capability: the substantive freedom to give form and value (through storytelling and all forms of cultural expression) to our experiences of being-in-relation with each other, with the world, and with its possibilities..

In our work with CPP, we suggested the need for forms of 'ecological leadership', committed to 'holding open spaces and structures' (Gross and Wilson 2019). Building on that suggestion, here we go further by arguing that understanding and supporting cultural infrastructure is essential to realizing the normative commitments of cultural democracy. In practice, this will involve developing strategic plans and specific actions that deliberately promote 'open systems and structures'—including relational goods, such as trust—in support of cultural capability, doing so through a commitment to stewarding the system as a whole. Exercising ecological leadership of this kind, directed towards a system of diverse social and cultural infrastructures, policy makers can purposively support the many kinds of spaces (including schools, libraries, arts organizations, community centres, citizen assemblies and many more) through which relationships can be formed, stories told, and ongoing democratic practices developed.

It is here that we return directly to the GND, and its project of 'infrastructure revolution' (Rifkin 2019, p. 73), with two specific arguments. The first is to suggest that cultural policy could—and should—deliberately seek to cultivate conditions conducive to the formation of social mobilizations in support of the GND. It is good for cultural policy via prerequisites for funding from Arts Councils, for example—to move individual organizations towards 'greening' their practice. But interventions at that level of ambition, focusing only on reducing cultural organizations' carbon footprints, are extremely limited in their potential to effect change. Instead, if social movements and popular involvement will be crucial to the success or failure of the GND project (Aronoff et al. 2019; Klein 2019; Pettifor 2019; Rifkin

2019), cultural policy makers should take up the challenge of cultivating conditions conducive to building local, regional, national and international movements able to develop and sustain momentum behind the GND.

Secondly, cultural policy needs to address what the 'democratic infrastructure' of the future should look like. If, as its advocates propose, the GND needs to be a *devolved* infrastructural revolution—with participatory decision-making taking place at regional and local levels across highly-networked energy infrastructures—what role can cultural policy play in creating conditions in which democracy can work well? Democracy cannot flourish if it is based only on occasional voting practices. Votes, and other modes of decision-making, need to be embedded within sustained conditions of relationship, discussion and deliberation. The greater the possibility of these conditions, the more trust there will be in the democratic process: with more confidence in the system of decision-making, and decisions increasingly conditioned by people's faith in each other. We suggest that, as part of the overall project of a GND, and in an age of radical uncertainty, cultural policy can take it upon itself to cultivate conditions of trust through its support for cultural infrastructures in which people develop relationships, tell stories, and know that their actions matter—conditions of hope (Gross 2019). Cultural infrastructure, as we characterize it here, is both the outcome of, and condition for, the development of relational goods such as trust and hope. Such relational goods, many of which are also fundamentally *public* goods, will be crucially important over the coming years.

In our 'ecological' approach to cultural democracy, the publicly funded arts, profit-making creative industries and everyday creativity are understood to be interconnected and interdependent. Any future GND will be a form of mixed economy, but there will be a major shift towards the expansion of the public realm, and the roll back of market mechanisms. To make this work, a massive effort will be required to reclaim the value of public goods, and the very idea of 'the public sphere' (McGuigan 1996). We suggest that cultural policy has a specific and vitally important role to play here. For it is precisely through our experiences of being-in-relation—that is, our 'aesthetic' experiences (Wilson 2020)—that people become aware of, and potentially more attached to, the forms of *relationality* in which they are embedded, and which enable their flourishing. It is in the development and diversification of cultural infrastructure, supporting such aesthetic experience, that cultural policy can promote the conditions necessary for a GND to garner popular support of the breadth, depth and energy required to take on the entrenched interests and powerful voices that will resist it.

11.4 A New Story of Cultural Policy

Latour (2018, p. 6) characterizes the present moment as one in which 'the migratory crisis has been generalized.' We are all, he suggests, now facing 'the ordeal of being *deprived of land*'. Latour's description of the present indicates the way in which the climate crisis is both a material phenomenon of an incontrovertible kind, and a powerful experience of human disturbance. The 'sense of vertigo, almost panic', he writes, 'that traverses all contemporary politics arises owing to the fact that the ground is giving way beneath everyone's feet at once, as if we all felt attacked everywhere, in our habits and in our possessions' (Ibid., p. 8).

For Latour (2018, p. 44)—as for others such as Haraway (2016), Monbiot (2017) and Gross (2019)—'we can no longer tell ourselves the same old stories.' Within these conditions, the GND offers the kind of new narrative required: with the potential, at least, to take on neoliberal hegemony and win, expanding the boundaries of political possibility. The need for 'new stories' is finding increasingly popular articulations—not reserved for academic analysts of hegemony. A recent book by journalist and former private equity manager Nesrine Malik (2019), *We Need New Stories: Challenging the Toxic Myths Behind Our Age of Discontent*, takes Brexit and Trump as a wake-up call, requiring new attention be paid to the 'myths' we live by. The use of the language of myths, however, slips too easily into a sharp separation between those stories which are true and those that are not. Challenging toxic untruths is, of course, an urgent task. But it is also vital to recognize, if projects of progressive change are to succeed, that narrative-making is an essential component of all political contestation (Mouffe 2019) and political imagination (Gross 2019).

In this context, it is striking to note Klein's use of the language of storytelling in her most recent writing on climate change. Notwithstanding the broad ambition of *This Changes Everything: Capitalism Vs the Climate* (Klein 2015), in *On Fire: The Burning Case for the Green New Deal* (Klein 2019) she makes the case for a new political story. Hope is 'the stuff of the political imagination' (Gross 2019), and for Klein, the matter of hope (and its opposites) is crucial to keep in mind when developing a rhetoric in support of the GND. 'One warning from the 1930s and '40s we would be wise to remember', she writes, 'is that when systematic crises cause political and ideological vacuums to open up, as they have today, it is not only humane and hopeful ideas like the Green New Deal that find Oxygen. Violent and hateful ideas do, too.' (Klein 2019, p. 40) It is precisely because we find ourselves in a 'populist moment' (Mouffe 2019), an 'interregnum' in which the old is dying and the new cannot be born (Fraser 2019, following Gramsci), that the need for a big political narrative—such as a GND—is essential. The alternative may well be a different story, one whose ending is unlikely to be a happy one.

Advocates of the GND, therefore, need to communicate effectively how it will make life *better*. There is much to be said about what the subplots of this new story should be, including, for example: new accounts of the good life (Oakley and Ward 2018; Skidelsky and Skidelsky 2012); a 4-day working week (Klein 2015); green jobs and a universal job guarantee (Aronoff et al. 2019); more accessible alternatives to fast and carbon-intensive consumption (Jackson 2017; Klein 2019); and universal health care (Aronoff et al. 2019). Part of the importance of many of these proposals is that they potentially help enable new approaches to and experiences of *social security*—which will be vital not only to people's physical wellbeing, but also to cultivating relational goods, such as trust and hope, as the disruptions caused by climate change develop. As Klein suggests, 'that's why putting universal health care in the Green New Deal is not an opportunistic add-on—

it's an essential part of how we will keep our humanity in the stormy future ahead' (Klein 2019, p. 268).

Our argument is that establishing the GND as a politically effective story requires the development and diversification of cultural infrastructure. This will have a crucial role in creating the conditions necessary for the vibrant practices of *devolved decision-making* and *social power* that champions of a radical GND make clear that it requires (Aronoff et al. 2019; Klein 2019; Rifkin 2019). We are arguing for the promotion of ecological leadership within cultural policy, committed to stewarding interconnected systems of cultural infrastructure in which opportunities for experiences of being-in-relation and storytelling are multiplied and democratized. These cannot be 'value neutral' spaces. They need to be underpinned by a range of normative commitments. This includes an explicit commitment to the GND itself. But also to the promotion of cultural democracy (Wilson et al. 2017; Gross and Wilson 2018), an ethics of care (Wilson and Gross 2017; Tronto 1993, 2013), and the cultivation of trust (Gross 2019).

Which countries, locations and policy makers are we speaking to here? Much of our research to date has focused, in the first instance, on the UK, whilst making conceptual contributions that we intend to be widely pertinent. Our current research project is supported by EU funds and focuses on ten cities across Europe (Gross et al. 2019). The arguments in this chapter are relevant to policy makers in different locations, and working at a variety of scales, from the intergovernmental to the municipal. One of the questions that an ecological approach to cultural policy raises is: *what are the scales that matter*, what are the spaces that become places—the locations and relationships that have meaning and consequence for people in giving form and value to their experiences of being-in-relation? (Gross and Wilson 2019) We therefore share these ideas with the hope that they be taken up by policy makers in many locations, and in many positions within cultural eco-systems. Moreover, we take it that those involved in cultural policy making are—and could be—much more numerous and heterogenous than only those who work for ministries of culture and arts councils.

At the same time, we appreciate the argument that some countries, given their power and position within the carbon-intensive economy, have a particularly important role to play in taking a lead on climate policies. For example, as Aronoff et al. (2019, p. 10) suggest, 'if the United States joined (and prodded) Europe and China in prioritising climate-friendly green investment, more than half the global economy would be invested in climate action.' A similar claim can be made for cultural policy. Based in the UK, we are aware of the international influence that this country's cultural policies have had, not least via the discourse of the creative industries (Gross 2020). Notwithstanding the far smaller size of its economy and, following Brexit, its potential marginalization, the adoption of a commitment to the GND as an overarching strategic goal of UK cultural policy—with a programme of radical and highly visible actions—could have an outsized influence. Given the very uneven visibility of 'cultural policy' as a policy domain internationally, it may, indeed, be important to use the language of creative industries or creative economy in seeking to realize the goals we are proposing here: a radical 'greening' that not only involves the reduction of emissions, but also names and supports the forms of cultural infrastructure that will be necessary to making a GND politically possible. Put more boldly, this is the time for cultural policy makers to realize their potential influence over conditions which will be central to the success of any GND, and to the future of human flourishing—namely, the infrastructures that enable people's democratic and creative connections with each other, and with the world.

11.5 Conclusion

To meet the challenges of climate change, infrastructural revolution is essential—and this is what the GND proposes (Aronoff et al. 2019; Klein 2019; Pettifor 2019; Rifkin 2019). This chapter has made the case for the forms of *cultural infrastructure* that will be necessary to achieving the GND and its aims—in addition to the headline transformations in energy, transport and communication systems—and for the role of cultural policy in supporting such infrastructures. For Klein, 'the Green New Deal will necessarily be a work in progress, one that is only as robust as the social movements, unions, scientists, and local communities that are pushing for it to live up to its promise' (Klein 2019, p. 266). Here we have argued that cultural policy must recognize this, and realize its own distinctive potential to promote the conditions required for developing and sustaining momentum behind a GND.

Cultural infrastructures are essential to building the conditions of trust, creative self-narration, and democratic action without which the GND will not be possible. On the one hand, we need a new shared story of where we are going. The discourse of the GND promises to offer this. On the other hand, we also need cultural policies that support many and diverse spaces for cultural capability to be exercised and developed, and from which many new stories can emerge: spaces that are conducive to ongoing, democratic processes of self-narration, deliberation, decision-making and action.

Connecting cultural policy to the provocations and potentials of the GND, as this chapter has begun to do, entails a future-focused remembering, recalling the possibilities of large-scale public action to effect systemic change. But these discussions of the GND have also helped us to make visible, in a distinctive way, the need to *stay with the trouble* (Haraway 2016): recognizing our individual and collective vulnerabilities, paying better attention to these conditions, and creating the new forms of public space and public goods with which to 'liv[e] *well* with the risks that inevitably attend human existence'—in other words, to live *courageously* (Lear 2008, p. 121). There is so much we need to re-future. Cultural policy may appear a parochial concern. But it has a specific, necessary and consequential part to play. Cultural policy makers of many kinds, in many locations, can foster conditions for courage by recognizing their own role in the GND, this 'grand story [...] to repair our relationship with the earth and with one another' (Klein 2019, p. 270). Recognizing this role—and making it central to all they do—will, in itself, be an ongoing act of courage.

References

Aronoff, K., A. Battistoni, D.A. Cohen, and T. Riofrancos. 2019. *A planet to win: Why we need a green new deal*. London: Verso.

Banks, M. 2018. Creative economies of tomorrow? Limits to growth and the uncertain future. *Cultural Trends* 27 (5): 367–380. https://doi.org/10.1080/09548963.2018.1534720.

Barker, V. 2019. The democratic development potential of a cultural ecosystem approach. *The Journal of Law, Social Justice and Global Development* 24: 86–99. https://doi.org/10.31273/LGD.2019.2405.

Bell, D., and L. Orozco. 2020. Neighbourhood arts spaces in place: Cultural infrastructure and participation on the outskirts of the creative city. *International Journal of Cultural Policy*. https://doi.org/10.1080/10286632.2019.1709059

Donati, P., and M. Archer. 2015. *The relational subject*. Cambridge: Cambridge University Press.

Fraser, N. 2019. *The old is dying and the new cannot be born*. London: Verso.

Gross, J. 2019. Practices of hope: Care, narrative and cultural democracy. *International Journal of Cultural Policy*. https://doi.org/10.1080/10286632.2019.1702032

———. 2020. *The Birth of the creative industries revisited: An oral history of the 1998 DCMS mapping document*. London: King's College London. https://doi.org/10.18742/pub01-017

Gross, J., and N. Wilson. 2018. Cultural democracy: An ecological and capabilities approach. *International Journal of Cultural Policy, 26*, 3, 328–343. https://doi.org/10.1080/10286632.2018.1538363

Gross, J., and N. Wilson. 2019. Creating the environment: The cultural eco-systems of creative people and places. *Creative People and Places*. Retrieved 13 Mar, 2020, from https://www.creativepeopleplaces.org.uk/our-learning/creating-environment

Gross, J., R. Comunian, B. Conor, T. Dent, J. Heinonen, U. Hytti, K. Hytönen, T. Pukkinen, P. Stenholm, and N. Wilson. 2019. *DISCE case study framework*. London: DISCE Publications.

Hallam, R. 2020. *Common sense for the 21st century. Only nonviolent rebellion can now stop climate breakdown and social collapse*. Version 0.3. Retrieved 13 Mar, 2020, from https://www.rogerhallam.com/wp-content/uploads/2019/08/Common-Sense-for-the-21st-Century_by-Roger-Hallam-Download-version.pdf

Haraway, D. 2016. *Staying with the trouble: Making kin in the Cthulucene*. Durham: Duke University Press.

Holden, J. 2015. *The ecology of culture*. Swindon: Arts & Humanities Research Council.

———. 2016. *Organism not mechanism: An ecological approach to cultural learning*. London: A New Direction.

Howkins, J. 2010. *Creative ecologies: Where thinking is a proper job*. New Brunswick: Transaction Publishers.

Jackson, T. 2017. *Prosperity without growth: Foundations for the economy of tomorrow*. 2nd ed. London: Routledge.

Jancovich, L. 2019. Creative people and places – an experiment in place-based funding. *Journal of Arts and Communities*, 9 (2): 129–147. https://doi.org/10.1386/jaac.9.2.129_1.

Jeffers, A., and G. Moriati. 2017. *Culture, democracy and the right to make art: The british community arts movement*. London: Bloomsbury.

Kelly, O. 1984. *Community, art and the state: Storming the citadels*. London: Comedia.

Klein, N. 2015. *This changes everything: Capitalism vs the climate*. London: Penguin.

———. 2017. *No is not enough: Defeating the new shock politics*. London: Penguin.

———. 2019. *On fire: The burning case for a green new deal*. London: Penguin.

Klinenberg, E. 2018. *Palaces for the people: How to build a more just and united society*. London: Penguin.

Labour Party. 2019. It's time for real change: The labour party manifesto. Retrieved 3 Mar 2020, from https://labour.org.uk/manifesto/

Latour, B. 2018. *Down to earth: Politics in the new climatic regime*. Cambridge: Polity Press.

Lear, J. 2008. *Radical hope: Ethics in the face of cultural destruction*. Cambridge: Harvard University Press.

Malik, N. 2019. *We need new stories: Challenging the toxic myths behind our age of discontent*. London: Weidenfeld & Nicholson.

Markusen, A., A. Gadwa, E. Barbour, and W. Beyers. 2011. *California's arts and cultural ecology*. Accessed 13 Mar, 2020, from http://annmarkusen.com/wp-content/uploads/2013/01/caarts-culture.pdf

Maxwell, R., and T. Miller. 2017. Greening cultural policy. *International Journal of Cultural Policy*, 23 (2): 174–185. https://doi.org/10.1080/10286632.2017.1280786.

McGuigan, J. 1996. *Culture and the public sphere*. London: Routledge.

Monbiot, G. 2017. *Out of the wreckage: A new politics for an age of crisis*. London: Verso.

Morgan, J., and E. Fullbrook. 2019. Introduction: Economics and civilization in ecological crisis. In *Economics and the ecosystem*, ed. E. Fullbrook and J. Morgan, 3–16. Bristol: World Economics Association Books.

Mouffe, C. 2019. *For a left populism*. London: Verso.

Nicholson, H., N. Holdsworth, and J. Milling. 2018. *The ecologies of amateur theatre*. London: Palgrave Macmillan.

Oakley, K., and J. Ward. 2018. The art of the good life: Culture and sustainable prosperity. *Cultural Trends*, 27:1, 4–17. https://doi.org/10.1080/09548963.2018.1415408

Pettifor, A. 2019. *The case for the green new deal*. London: Verso.

Rifkin, J. 2019. *The green new deal: Why the fossil fuel civilization will collapse by 2028, and the bold economic plan to save life on earth*. New York: St. Martin's Press.

Skidelsky, R., and E. Skidelsky. 2012. *How much is enough? Money and the good life*. London: Penguin.

Tronto, J.C. 1993. *Moral boundaries: A political argument for an ethic of care*. London: Routledge.

———. 2013. *Caring democracy: Markets, equality, and justice*. New York: New York University Press.

United Nations. 2015. *Paris agreement*. Paris: United Nations. Retrieved 6 Mar 2020, from https://unfccc.int/files/essential_background/convention/application/pdf/english_paris_agreement.pdf

Wilson, N. 2020. *The space that separates: A realist theory of art*. Abingdon: Routledge.

Wilson, N., and J. Gross. 2017. *Caring for cultural freedom: An ecological approach to supporting young people's cultural learning*. London: A New Direction.

Wilson, N., J. Gross, and A. Bull. 2017. *Towards cultural democracy: Promoting cultural capabilities for everyone*. London: King's College London.

Chapter 12
'Exhaustion' as a Driver of Change in the Creative Economy

Ealasaid Munro

12.1 Introduction

Since the late 1990s, British policymakers have identified the creative industries as a key driver of the national economy (O'Brien 2014). According to the Centre for Economics and Business Research the creative industries contributed £8.5 billion to the economy in 2016 (CEBR 2017, p. 4),[1] and creative-economic strategies have been mooted as solutions to stalled growth, un- and under-employment, moribund town and city centres and a whole range of other economic and social crises. Whilst the idea that creativity represents a 'fix' for these diverse crises has been widely critiqued, Schlesinger et al. (2015) have argued that the trajectory of creative-economic policy remains largely unaltered.

The fact that the creative-economic policy 'juggernaut' rumbles on in this way is a source of frustration for those of us who have questioned the foundations upon which the creative economy is built, unpicking its inconsistences and exploring its existing forms. There is now a growing body of critical work that seeks to 'rethink' the foundations and trajectory of the creative economy. As Oakley et al. (2018, p. 311) observe, 'the rationale for the creative economy [...] has reached its limits. Faced with critical environmental concerns, and increasing demands for justice in

[1] In the last 10 years there have been significant improvements in ICTs, leading to growth in digital creative production, distribution, and consumption. The established creative industries and newer 'digital industries' are often grouped together as a separate economic sector—the 'creative economy'.

The original version of this chapter was revised. The correction to this chapter is available at https://doi.org/10.1007/978-3-030-49384-4_12

E. Munro (✉)
Office of Communications (Ofcom), Glasgow, UK

K. Oakley, M. Banks (eds.), *Cultural Industries and the Environmental Crisis*,
https://doi.org/10.1007/978-3-030-49384-4_13

the ways we produce, consume and experience culture, it is vital that we now explore possible critical paths for its future.'

This chapter contributes to work that explores creatives workers' resistance to key tenets of creative-economic policy discourse, and their desire to produce alternative economic imaginaries through their work (see also Luckman 2018; McLean 2017; Munro 2016). Using Gilles Deleuze, I show how 'exhaustion' might be a driving force for change among creative workers. It draws on ethnographic material from two short research projects, including interviews with 23, predominantly rurally-based, creatives and participant observation in a range of settings.[2] While 'exhaustion' is my key theme, the analysis also contributes to wider debates about whether ideas on the creative economy based on urban experience can work in rural settings (Bell and Jayne 2010, p. 210; Harvey et al. 2012; Luckman 2012).[3]

The lives of the creative workers in my study are also shaped by two specific policy contexts. First, they work within a rural creative economy that has a patchwork quality (Bell and Jayne 2010; Roberts and Townsend 2016), a result of both poorly-developed infrastructure and attempts to combine 'urban' creative-economic policy frameworks with 'rural' policy interventions focused on the heritage sector and forms of creative practice such as craft. Creative workers must often develop the creative economy 'from below' to patch the gaps in the institutional infrastructure (Munro 2016). Second, although these workers are located in a Scottish policy context where devolution has offered a great degree of national control over cultural and creative industries development strategies and a separate national funding apparatus, the Scottish Government's creative industries development strategy is still broadly in line with Westminster (Schlesinger et al. 2015). However, rural Scotland has a complex institutional infrastructure in which sector-specific funding and support agencies overlap with geographically-bounded ones, a complexity that acts as a barrier for creatives seeking funding and support (Munro 2016; Schlesinger et al. 2015).

12.2 Discourses of Growth in the Creative Economy

One recent strand of scholarship explores the role of the creative economy in an era of stalled economic growth. As Banks (2018, p. 366) notes, 'official statistics show that in the decade since the financial crash the creative industries have recovered to become one of the UK's most dynamic industrial sectors, consistently reporting

[2] The chapter draws on material gathered as part of two research projects concerned with the support landscape for creative practitioners in Scotland. The first, 'Supporting Creative Business', was funded by the Arts and Humanities Research Council. The second, 'Towards a model of support for the rural creative industries', was funded by the University of Glasgow.

[3] See also the wealth of research on how creativity contributes to rural development in Australia (Argent et al. 2013; Gibson 2010; Gibson and Connell 2015; Waitt and Gibson 2013).

growth in excess of the national average.' Despite the gloomy outlook for the UK economy more generally—particularly in the context of Brexit—the creative economy is lauded as a success story.

Government narratives about the creative economy emphasize its importance to economic growth at a range of scales. Funding bodies and business support agencies also aim to bring about growth—in key creative sectors, and in creative businesses and individuals (Munro 2017; Schlesinger et al. 2015). The policy landscape is predicated on the idea that, through funding and support, creatives can develop their entrepreneurial skills and build a business from their talent. For Luckman (2018, p. 314), an entrepreneurial subjectivity is characteristic of the experience of working within the creative economy: 'the increasing absence of secure employment options in the neoliberal workplace demands greater individual responsibility for generating work of workers themselves.' In the absence of professional development frameworks, entrepreneurs internalize a commitment to professional and personal growth.

Banks (2018), Luckman (2018) and Oakley et al. (2018) have questioned whether the creative economy can be seen as a growth area when the economy more generally is understood to be stagnating. The crude nature of metrics designed to measure growth (such as GVA and GDP) are unsuitable for measuring the creative industries and their contributions to local, regional and national economies, and adherence to purely economistic conceptions of growth fails to account for the more-than-capitalist values that animate much creative production (McRobbie 2016). Furthermore, 'boosterist' discourses of growth mask the poor working conditions, extreme precarity, and tendency towards burnout inherent in the creative labour market (Hesmondhalgh and Baker 2013). Most importantly, in the context of impending climate catastrophe, growth—as it is generally understood by economists—is unsustainable, and policymakers and participants in the creative economy may be complicit in 'lighting the path of return to a world where dreams of unfettered economic expansion might once again be dreamt' (Banks 2018, p. 367).

The creative industries are often seen as a 'clean' route to growth. However, creative production is not environmentally benign: 'the growth of digital technology that it promotes is highly dependent on energy use and the association with lifestyle goods, via fashion, craft, design and advertising links it in many ways to conspicuous (over)consumption' (Oakley et al. 2018, p. 15). Failure to tease out the tensions between elements of the so-called 'triple bottom line' of the economic, social and environmental impacts of creative activities has led to a blind spot relating to creativity and the environment, and serious consideration of the environmental impact of the creative economy has yet to percolate into the policy mainstream.[4] Relatively

[4]Indeed, in Creative Scotland's (2016, p. 9) Creative Industries Strategy the 'triple bottom line' is identified as 'economic, social and cultural value' with no mention made of the environment. Creative Scotland, like all government bodies in Scotland, has a statutory requirement under the Climate Change (Scotland) Act 2009 to contribute to delivery of the Act's emissions reduction targets. Creative Scotland's Environment Policy (2013) is primarily concerned with the organization's own operation, although Creative Scotland has in recent years implemented a number of

little scholarship explores the contribution of the creative economy to climate change (something that this volume seeks to address), although there is a growing body of work that explores how creative labourers are rethinking their practice in light of an increased awareness of environmental issues (Anderson et al. 2016; Collins et al. 2018; Collins and Fahy 2011; Luckman 2018; Oakley et al. 2018). The twin tendencies identified in much of this literature—towards 'burnout' and environmental exhaustion—are rarely addressed concurrently. This chapter shows how the tactics that creative workers employ in order to ward off or manage a state of burnout are bound up with a range of other concerns, including the environment.

12.3 The Creative Economy: An Exhaustion 'Machine'?

In *Capitalist Realism*, Mark Fisher (2009) argues that the encroachment of capitalist dynamics into all areas of social life has led to our inability to imagine alternative economic and social futures. For Brigstocke (2016), who draws on Fisher's (2009) work our ability to engage productively with the future has been worn away by a sense of powerlessness and exhaustion in the face of the seemingly-unstoppable twin processes of capitalist expansion and environmental degradation. For these authors, the future has been 'occupied, colonized [...], destroyed' (Ibid., p. 20). In this work, there are echoes of Gilles Deleuze's (1996) philosophy of exhaustion.

Where Deleuzian theory has been used within creative-economic scholarship, it has normally been in relation to his work—in collaboration with Felix Guattari—on the assemblage, and the rhizomatic social formations that can disrupt or subvert established ways of 'doing' or 'thinking' the creative economy (see Daskalaki and Mould 2013). Assemblage theory allows us to theorize how the creative economy is a product of the interactions between multiple policies, institutions, intermediary agencies, creative businesses and individuals. This highlights a key tension at the heart of the creative-economic project between what Deleuze and Guattari (1988) call the 'arborescent' structure—the established policies, practices and modes of evaluation that constitute the creative economy—and possible 'lines of flight', attempts by artists, activists and critics to recover the progressive potential in creativity. The language of assemblage is appealing because it contains the possibility of change (Nail 2017) and it has been used productively to trace the development of social movements and changes within institutions and organizations. Deleuze also explicitly linked creativity to lines of flight: it is our desire to escape the status quo that leads us to create (Deleuze and Parnet 2007; for creative-industries specific applications see Duff and Sumartojo 2017; Lithgow and Wall 2018; Mar and Anderson 2010).

However, in his later work, Deleuze moved away from theorizing the productive nature of creativity. His short essay *L'epuise* (*The exhausted*) (Deleuze 1996)

---

initiatives relating to sustainability within the sectors they fund, including requiring funded organizations to report on their carbon footprints and sustainability initiatives, and offering advice and support to creative businesses on how to act more sustainably.

sketches out the contours of a state of complete exhaustion. For Deleuze, exhaustion is more than just a feeling of tiredness, or of being 'worn out', but instead a state from which it is impossible to recover: '[t]he tired has only exhausted realization, while the exhausted exhausts all of the possible. The tired can no longer realize, but the exhausted can no longer possibilitate' (1996, p. 3). This sees a fundamental 'closing of the infinite forces of creation and potentiality that [Deleuze] had been attempting to release in his earlier work' (Kerslake 1996, p. 113). Exhaustion is the state that is reached when it is evident that an arborescent system is no longer functioning, and where the formation of lines of flight have been arrested.

We are no longer able to ignore the fact that the creative economy contributes to the exhaustion of resources—human and environmental. Creative work is often extremely precarious and recent scholarship shows that long hours, low pay, and a sense of constantly having to 'hustle'—or juggle multiple income streams—can, over time, contribute to burnout (Hesmondhalgh and Baker 2013; Njaradi 2014). Within the creative economy, where much work is freelance or project-based, this can be compounded by a lack of opportunities for further training and career development, and for breaks for illness, child-rearing or other caring responsibilities (Siebert and Wilson 2013). These conditions also limit creative workers' ability to plan for the future, leading to feelings of paralysis (Schlesinger et al. 2015) or, in Deleuzian terms, leaving them unable to 'possibilitate'.

It may be becoming harder to 'possibilitate', to recover the progressive potential in the creative economy, but it is not altogether *impossible* (see Mould 2018). In this chapter, I reconfigure Deleuze to show how both the threat of exhaustion—and the feeling or state of it—can be generative and provide the impetus to 'do' creative work differently. My analysis highlights how creative workers' articulate the human and environmental cost of working in the creative economy. If the idea of exhaustion is already a key theme in scholarship on the experience of working within the creative economy, the concept of 'exhaustion' also animates much critical work on the creative and media industries' impact on the environment (Christensen and Nilsson 2018; Kääpä 2018; Maxwell this volume). In my research, creative workers occasionally elided bodily and environmental exhaustion, and saw reshaping their practice as key to managing both of these issues. My aim is to show how these twin concerns can give rise to lines of flight, animating creative production and creative workers' drive to rethink their practice and their orientation towards established ways of working in the creative sector.

12.4 Escape to the Country? Rurality as a Line of Flight

The creative economy is characterized by portfolio working, extreme precarity, high rates of burnout, and of physical and mental exhaustion (Hesmondhalgh and Baker 2013). For a number of my interviewees, choosing to relocate to a rural location (or in some cases, to return home after years spent training and working away) stemmed from the desire to ward off, or respond to, a state of exhaustion (see also Gibson 2010; Luckman 2012; Waitt and Gibson 2013).

One craftsperson narrated her 'moves' around the country over the last decade. She had moved from her small hometown in the central belt of Scotland to Edinburgh to complete her degree. Upon completion, she moved to London—'a common move', she told me, among young graduates—to gain experience through internships, and to take advantage of the concentration of creative workers, businesses and clients in London. After a few years of 'the grind', and sensing burnout, she applied for a series of residencies in rural Scotland, eventually taking up an opportunity in the Northern Isles. She told me: 'my mental health improved almost overnight'. She cited the slower pace of life, the close-knit community of artists (and non-artists) that she had slotted into, and the 'nourishing' effect of the natural landscape as important factors in her decision to relocate there permanently. She cautioned that relocation wouldn't be for everyone but for her, it worked, and her creative practice and her orientation towards work had changed as a result. She was concerned with keeping her business manageable and protecting her work-life balance. She saw this as vital in order to protect her mental health and to reduce the potential for exhaustion, but also in order to ensure her enthusiasm for her practice wasn't dimmed.

This craftsperson's testimony was echoed by other interviewees, who also saw a move to the country as a way of dealing with exhaustion, or the threat of it. One textile artist, originally from Moray, told me that, after coming to the realization that she was 'worn out', she returned home to take stock:

> I was living in London, I'd done internships, I was doing really well as far as it went. I was selling. But I was barely making enough to live on, commuting hours a day, working in a shop, no social life. Just worn out, and I thought 'Fuck it'. The village'd got broadband, mum and dad still had my old room, so I went home for a bit and [re-thought] what I was doing.

There are a number of reasons that craftspeople may choose to base themselves in a rural area rather than a city (Luckman 2012). For my interviewees, advantages of a rural location included cheaper rents or lower property prices relative to cities; more space; a slower pace of life; and a feeling of being embedded within and valued by their communities.

One key problem for many of the creatives that I spoke to was that many intermediaries saw a commitment to 'growth' as a precondition for funding and support. Highlands and Islands Enterprise (HIE)—a key funder for businesses in the Highlands and Islands, and for many of my interviewees—states in its cultural policy that '[a]spects of regional culture of importance to HIE are those which have the greatest potential to realise economic and social growth' (HIE 2013, p. 6). It was felt that despite HIE's commitment to supporting businesses that contributed to economic and social growth, economic growth was afforded primacy. Given the economic, social, cultural and environmental fragility of rural Scotland, this narrow focus on economic growth was seen as problematic and contributing to the exhaustion of resources—both human *and* environmental.

One designer summed up her reasons for 'staying small', explaining:

> The pressure on small businesses is always more, more, bigger, bigger [...] I would like to sell more, yeah. But if I expand much more I will need to go to China for manufacture, maybe Turkey, Romania. That is going to be a lot more work for me. If I stay small, I can

keep my production in the UK, as local as possible, that's jobs here, shorter supply chain, probably less waste, less hassle.

For this designer, to commit to unrealistic or undesirable expectations of growth was to run the risk of exhaustion. However, she felt that intermediaries such as funders and support agencies 'weren't interested' if creative workers' ambitions were to stay small (see also Schlesinger et al. 2015). This designer's testimony shows an elision between her desire to ward off exhaustion by cultivating a manageable work-life balance, and her concern about her business's environmental impact: for her, 'staying small' is one way of dealing with both pressures. For these respondents, exhaustion had given rise to both a literal line of flight (a move to the country) and a figurative one (often expressed as a desire to 'stay small', or more generally to conceive of ways of working within the creative economy that privileged a healthy work-life balance, had a social or community-building function, and that safeguarded their physical and mental health).

A number of my interviewees felt that one way of responding to exhaustion was to opt-out of what one arts manager called 'funded culture', because of the expectations of business growth that came from funders such as Creative Scotland or HIE. Being part of funded culture often meant a loss of autonomy, something that Hesmondhalgh and Baker (2013) found was essential to people's conception of what constitutes 'good work' in the cultural industries. For some of my respondents, engaging with funded culture meant limiting creativity, a closing-off of creative lines of flight because of the need to align projects with institutional priorities and key performance indicators (KPIs), and submit to economistic modes of assessment and evaluation. This arts manager told me, 'I didn't want to take [HIE's] money unless I had to because they would scrutinize the accounts, there would be so many targets to meet and reports to file.' This interviewee went on to say that taking money from funding agencies limited his ability to be responsive when coming up with new initiatives because of the inflexible 'input-activity-outcome' structure of many funded projects.

The creative-economic assemblage constructs 'linear, panoptic power structures' that subjugate creative workers, robbing them of their sense of autonomy (Lithgow and Wall 2018, p. 355; see also Holt and Lapenta 2010; Huws 2010). To return to the language of Deleuze, the prescriptive and panoptic nature of the 'arborescent' institutional funding and support infrastructure was seen by some respondents as a contributing factor in their exhaustion, or as introducing the threat of exhaustion. One photographer told me that he was considering giving up his practice, citing burnout: 'The whole sector is fucked [...] It's set up to profit off overwork, and people end up over-promising [in funding applications] and then either killing themselves to do what they said they'd do, or just [...] not getting it done.'

For this photographer, it was almost impossible to make a living in the creative economy without engaging with the institutional funding landscape. However, he felt that the way that the sector is currently organized undervalues labour and encourages overwork. For him, the institutional infrastructure of the creative economy—and the working practices it propagates—explicitly contributes to the exhaustion of the workforce.

12.5 Emerging Narratives of Environmentalism, or Entrepreneurialism Redux?

Thus far this chapter has highlighted how exhaustion was invoked by interviewees as the catalyst for changing their creative practice, and in some cases rethinking their orientation towards creative labour more generally. Talk of personal and environmental exhaustion was sometimes elided in interviewees' testimony: where individuals spoke of changing their practice in order to ward off or manage exhaustion, they may also invoke narratives of environmental concern. Creative workers' desire to 'go slow', or 'stay small' was seen as a way of responding to exhaustion whilst also reducing the environmental impact of their work (see also Luckman 2018).

Research by industry bodies has shown improving awareness of environmental issues within the creative sector in Scotland. For example, Creative Scotland requires all Regularly Funded Organizations to report carbon emissions, and a recent report by Creative Carbon Scotland (2017, p. 2) notes increased engagement with environmental issues, particularly around recycling and waste and the reduction of carbon emissions. Whilst creative businesses demonstrated a good understanding of environmental issues, they faced barriers when attempting to reduce their impact (Ibid.). My interviewees found the language of 'carbon emissions', and practicalities around monitoring and reducing emissions, were key barriers. These measures do not form part of creatives' training, even for those who had recently left art school or university. The onus for drawing down training in this area, reducing environmental impact, and quantifying this in line with intermediaries' reporting structures has come to fall predominantly on the shoulders of individuals.

One arts manager told me that she felt that creative workers had been egregiously slow to 'wake up' to their environmental impact. For her, the pressures of day-to-day work within the creative industries had, in her experience, long displaced concerns about the wider impact of creative production. However, she felt that growing concern over environmental exhaustion was driving change:

> Artists are so worried about putting food on the table that they don't think about the environment [...] I think things are changing though, especially here, where people come to be [...] inspired by [the landscape]. Maybe they're primed in a way to think, 'Well what if this wasn't here?' or 'What if this was different?'

Among the creatives I spoke to, many had begun to introduce measures to reduce their environmental impact, aware of the role that their work might have to play in the exhaustion of resources. These included having a good understanding of supply chains and keeping them as short as possible; being aware of energy consumption and attempting to reduce it; use of renewables; use of natural or biodegradable materials for manufacture and packaging; ensuring products were recyclable; making products that had lasting using or appeal rather than a short shelf-life; ensuring products could be repaired rather than thrown away; and reducing travel.

The intertwining of creative work with individuals' lifestyles and sense of 'self' is well-recognized (Duffy 2016; Taylor and Littleton 2016). When narrating what they did to reduce their environmental impact, interviewees often conflated work-related initiatives with lifestyle changes. The photographer quoted above listed his attempts to reduce his environmental impact in a WhatsApp conversation,[5] some time after his interview:

> Photography relies on extraction in the manufacture of cameras, SD cards, film, and in development, printing etc. This is bad, but I do what I can
> (2nd hand cameras, repairing if possible, reducing waste)
> I recycle
> I avoid single-use plastics
> I'm trying to be vegan (not always successfully)
> I don't really fly and if I do I will carbon offset.

When visiting a fashion designer at her studio, I asked her what she was doing to reduce her environmental impact, and again, the intertwining of work and lifestyle changes was evident. Despite beginning to trade internationally, she was determined to stick with her current suppliers and factories: she uses local wool, machine knitted, and finished in the studio. This system kept her supply chain manageable. The combination of machine-knitting and hand-finishing meant she could exploit economies of scale but her products retained a crafted 'feel'. Whilst she recognied that her products were luxury goods, she felt that producing one-off pieces that were built to last was better than producing fast fashion. She travelled a lot for work, but had cut down significantly, using Skype and other digital solutions where possible. As with many rural creatives, her studio was attached to her croft house, hence many of her initiatives to reduce her environmental impact were shared between her 'home' space and her 'work' space. On the one hand this made living with a smaller environmental footprint easier, as it became 'second nature' to reduce consumption, but on the other it was hard to accurately track things like energy usage because it was impossible to separate work from everyday domestic life.

Given the way that creative labour is bound up with individuals' identities, lifestyles, and home-spaces, the commingling of work-related initiatives and lifestyle changes should come as no surprise. However this means that measures to reduce environmental impact within the creative economy remain largely individualized. My data suggests that creatives' entrepreneurial subjectivities have simply been 'expanded' in order to encompass these issues. In this way, environmental stewardship measures have become rescaled, with accountability shifted onto individuals. However 'lifestyle' activism of the type outlined here is recognized as largely ineffective in tackling climate change, in the face of continued capitalist expansion and the attendant exhaustion of natural resources. Furthermore, work-related and lifestyle changes of this kind are often the preserve of the (relatively) privileged, and out of reach of many who may be struggling. Only some creative workers are able to 'resist—generally from positions of class and racial privilege—the entrepreneurial imperatives otherwise presumed of idealised neoliberal subjects' (Luckman 2018, p. 317).

[5] This has been edited for clarity.

The creative industries have a soft power, their ability to influence others through their communication of ideas (Leach 2016). This influence is one distinctive contribution that creatives might make to addressing environmental issues. As Boykoff et al. (2010, p. 2) argue in relation to media representations, 'cultural politics of climate change—the production and consumption of a multiplicity of climate change media and media forms— is as crucial to getting climate change "work" done as the climate science itself.' Many of my interviewees saw themselves as raising awareness, both through their creative practice and through dialogue, lobbying, and helping others to solve problems, moving away from individualized approaches towards collective responsibility.

One visual artist from North West Scotland had shifted her practice to focus on landscapes: 'People come here to buy and we always have a conversation about the piece. My latest collection is all seascapes, so I talk to them about the sea, about fishing, warming, changes I've seen myself.' She saw herself as practising consciousness-raising through her art. Artists can play a role in communicating climate change science (Doyle 2016; Nurmis 2016), but most research has focused on specific artworks or exhibitions rather than how artists' or creative workers express their values in a range of activities ancillary to their practice. For example, my research shows that elements of creative workers' branding could be seen as having a similar consciousness-raising aspect. At an industry event run by HIE in July 2015, a panel of creatives spoke of how they exploited the possibilities of improved ICTs in order to offset the problems of working from rural areas. A designer on the panel said that ethical consumption was becoming increasingly important to consumers, and that this was a market she was trying to tap into through emphasising her 'slow' approach to fashion. Astute use of social media allowed her to showcase the provenance of her products. As she spoke, I checked her Twitter and Instagram, as well as the accounts of other designers she was linked to; a large part of their 'advertising' through these channels entailed giving followers an insight into life on the islands. The visual nature of these media allowed them to create 'virtual studios' (Budge 2013) and enabled them to document how local histories of making had influenced their practice, and how a rural location had influenced the aesthetic of their work. This had the effect of conveying a sense of the importance of craft, history, and the natural environment.

Digital 'solutions' of this kind are not always as 'environmentally-friendly' as they appear; they also contribute to resource extraction and exhaustion (Kääpä 2018). Greenpeace (2014) estimates that the ICT sector now comprises 2% of global emissions. Whilst using Skype might, on balance, be better than flying to a meeting, the environmental impact of ICTs was routinely underestimated by my respondents.

Creative workers often see themselves as having responsibilities towards their communities, particularly where economic, social, and environmental systems are fragile. An arts manager spoke of how concern about environmental degradation and planetary exhaustion had led members of the local creative community to instigate island-wide measures to reduce their environmental impact. The following is adapted from field notes:

In her interview, this arts manager had told me: 'A lot of us have come here for the same reason, to reclaim a work-life balance [...] Here we can put back into the community [more easily]. It's about keeping this place going.' After spending some time shadowing her, and speaking to other residents, I saw examples of this community-wide approach in action. The island is already self-sustaining in terms of energy, with all households and businesses powered by renewables. However the community was trying to do more to reduce its environmental impact. My interviewee had spoken about how the island was attempting to streamline suppliers (even doing a community-wide online shop, rather than a number of individual ones), band together to recycle waste (especially where specialist recycling facilities were needed) and increasing use of car-shares and other forms of community travel.

In this arts manager's testimony—which, taken as a whole, represented one of the most nuanced, complexly wide-ranging examinations of 'exhaustion' that I came across—we can see how creative labourers are beginning to 'de-individualize' concern for the environment, to push back against the expansion of entrepreneurial subjectivities to encompass environmental issues. The manner in which the island banded together around the huge, multi-faceted issue of climate change was heavily influenced by its distinctive community history. On one level, their community-wide initiatives were important because individual responses were seen as largely inadequate to tackling the issue of planetary exhaustion, with more radical solutions needed. One another level, however, 'de-individualizing' responses to pressing issues was seen as a way to ward off exhaustion on the part of individuals who may already be struggling with the burden of forced entrepreneurship (Oakley 2014), or the unmanageable expansion of those entrepreneurial subjectivities that are increasingly necessary for survival within the creative economy.

12.6 Conclusions

We can no longer deny the fact that creative work contributes to the exhaustion of resources, both human and environmental. A profound anxiety about 'exhaustion'—the threat of it, and the feeling or state itself—was framed as a catalyst for change by creative workers in my study. These wide-ranging ideas about 'exhaustion' can usefully contribute to a range of research areas. This includes scholarship that explores working conditions within the creative economy and the development of entrepreneurial subjectivities (Gill 2014; Oakley 2014; Scharff 2016); that examines the role of the creative economy in relation to discourses of growth (Banks 2018; Luckman 2018; Oakley et al. 2018); and that has begun to explore creatives' growing awareness of the environmental impact of their work.

I have employed a tempered version of Deleuze's (1996) metaphor of 'exhaustion' to frame creatives' orientation towards the creative-economic 'assemblage'. Many of the creatives I spoke to found the experience of working within the creative economy profoundly exhausting, characterized by long hours, low pay, precarity and a frustrating lack of alignment between their own principles and those which animate the funding and support landscape. Interviewees developed several 'tactics' for responding to, or warding off, exhaustion including staying small, and increasing their job satisfaction by incorporating broad-based economic, social, cultural and environmental concerns into their business models rather than prioritizing growth. Several of my interviewees had made the conscious decision to opt out of what they called 'funded culture', citing unreasonable expectations around growth as one reason for doing so.

Experts have warned that to avoid climate catastrophe, urgent action needs to be taken both in terms of limiting emissions and curbing resource extraction and environmental degradation. Whilst it is increasingly recognized that the industrial economy as we know it needs root and branch reform, and that continued capitalist expansion is not sustainable, national governments are unwilling to abandon the idea of growth. Hence a politically expedient portrait of the creative economy—as a key route to 'clean' growth—has emerged. However, those who champion the creative economy in this regard conveniently ignore the contribution that the creative, cultural and digital industries make to emissions, resource extraction and environmental degradation. As Banks (2018, p. 369) has written,

> Creative, cultural and digital industries [...] are highly energy-intensive and often egregiously polluting; this includes many of the world's leading digital technology and communications companies, the global film and television industries, publishing, music and the transport, circulation and logistical systems that sustain them.

Whilst many of my respondents expressed concerned but often contradictory views about the impact of their work on the environment, anxieties about exhaustion are undoubtedly driving environmental activism on the part of creative workers. Interviewees had incorporated a series of stewardship measures into business practice, including monitoring supply chains, monitoring and reducing emissions, waste management, and creative use of ICTs to reduce travel. Given the way that creative labour is bound up with individuals' identities, lifestyles and home-spaces, work-related initiatives and lifestyle changes were often inextricably linked. Creatives also took on responsibility for consciousness-raising around environmental issues and coming up with collective measures to reduce their impact on the environment. However, given the structural composition of the creative economy—where most businesses are microbusinesses or sole traders—this drive intersects with discourses of creative entrepreneurialism in a way that rescales the issue of climate change. This amounts to a process of responsibilization 'in which climate change becomes the responsibility of the individual rather than of governments or regulators who might affect significant policy changes through altering production and distribution' (Boykoff et al. 2010). In this case, this rescaling may work to further absolve governments, regulators and intermediaries from their responsibility to intervene at the macro-scale to arrest damaging industrial practices.

I am not suggesting that the actions of a small set of creative workers is going to alter the trajectory of global capitalism. However, this chapter shows that awareness of, and commitment to, alternative ways of 'doing' creative work is gathering pace, driven by concerns about the exhaustion of human and environmental resources. Creative workers' commitment to thinking outside of established economic models, and to privileging ways of working animated by principles of care rather than 'growth', has clear policy implications. It is imperative that policymakers and intermediaries rethink the assumptions that underpin the current creative-economic assemblage, and the kinds of funding and business support that act as regulatory structures for the development of the creative economy.

References

Anderson, A.R., C. Wallace, and L. Townsend. 2016. Great expectations or small country living? Enabling small rural creative businesses with ICT. *Sociologia Ruralis* 56 (3): 450–468.

Argent, N., M. Tonts, R. Jones, and J. Holmes. 2013. A creativity-led rural renaissance? Amenity-led migration, the creative turn and the uneven development of rural Australia. *Applied Geography* 44: 88–98.

———. 2018. Creative economies of tomorrow? Limits to growth and the uncertain future. *Cultural Trends* 27 (5): 367–380.

Bell, D., and M. Jayne. 2010. The creative countryside: Policy and practice in the UK rural cultural economy. *Journal of Rural Studies* 26 (3): 209–218.

Boykoff, M., M. Goodman, and J. Littler. 2010. *Charismatic megafauna? The growing power of celebrities and pop culture in climate change campaigns*, Environment, politics and development working paper series. London: Department of Geography, King's College.

Brigstocke, J. 2016. Exhausted futures. *GeoHumanities* 2 (1): 92–101.

Budge, K. 2013. Virtual studio practices: Visual artists, social media and creativity. *Journal of Science and Technology of the Arts* 5 (1): 15–23.

CEBR. 2017. *The contribution of the arts and culture to the national economy*. London: CEBR.

Christensen, M., and A.E. Nilsson. 2018. Media, communication, and the environment in precarious times. *Journal of Communication* 68 (2): 267–277.

Collins, P., and F. Fahy. 2011. Culture and creativity: A case study from the west of Ireland. *Cities* 28 (1): 28–35.

Collins, P., M. Mahon, and A. Murtagh. 2018. Creative industries and the creative economy of the west of Ireland: Evidence of sustainable change? *Creative Industries Journal* 11 (1): 70–86.

Creative Carbon Scotland. 2017. *Scoping study on carbon reduction strategies for the craft and digital content sectors of the arts and creative industries in Scotland*. Edinburgh: CCS.

Daskalaki, M., and O. Mould. 2013. Beyond urban subcultures: Urban subversions as rhizomatic social formations. *International Journal of Urban and Regional Research* 37 (1): 1–18.

Deleuze, G. 1996. The exhausted. *Parallax* 2 (2): 113–135.

Deleuze, G., and F. Guattari. 1988. *A thousand plateaus: Capitalism and schizophrenia*. London: Bloomsbury Publishing.

Deleuze, G., and C. Parnet. 2007. *Dialogues II*. New York: Columbia University Press.

Doyle, J. 2016. *Mediating climate change*. London: Routledge.

Duff, C., and S. Sumartojo. 2017. Assemblages of creativity: Material practices in the creative economy. *Organization* 24 (3): 418–432.

Duffy, B.E. 2016. The romance of work: Gender and aspirational labour in the digital culture industries. *International Journal of Cultural Studies* 19 (4): 441–457.

Gibson, C. 2010. Guest editorial—Creative geographies: Tales from the 'margins'. *Australian Geographer* 41 (1): 1–10.

Gibson, C., and J. Connell. 2015. The role of festivals in drought-affected Australian communities. *Event Management* 19 (4): 445–459.

Gill, R. 2014. 'If you're struggling to survive day-to-day': Class optimism and contradiction in entrepreneurial discourse. *Organization* 21 (1): 50–67.

Greenpeace. 2014. *Clicking green: How companies are creating the green internet*. Greenpeace: Washington, DC.

Harvey, D.C., H. Hawkins, and N.J. Thomas. 2012. Thinking creative clusters beyond the city: People, places and networks. *Geoforum* 43 (3): 529–539.

Hesmondhalgh, D., and S. Baker. 2013. *Creative labour: Media work in three cultural industries*. London: Routledge.
HIE. 2013. *Ambitious for culture*. Lochgilphead: HIE.
Holt, F., and F. Lapenta. 2010. Introduction: Autonomy and creative labour. *Journal for Cultural Research* 14 (3): 223–229.
Huws, U. 2010. Expression and expropriation: The dialectics of autonomy and control in creative labour. *Ephemera: Theory and Politics in Organization*. http://www.ephemerajournal.org/contribution/expression-and-expropriation-dialectics-autonomy-and-control-creative-labour. Accessed 3 Feb 2020.
Kääpä, P. 2018. *Environmental management of the media: Policy, industry, practice*. London: Routledge.
Kerslake. 1996. Translator's Introduction. Parallax 2 (2): 113–135.
Leach, S. 2016. *Scientific imagining: studio based research into genre images of science and how art might interpret modern science*. Unpublished Phd thesis, Royal Melbourne Institute of Technology.
Lithgow, M., and K. Wall. 2018. When the dead rise: Encountering resistant legacies of creative economy within an artist's residency at a municipal cemetery. *Cultural Trends* 27 (5): 353–366.
Luckman, S. 2012. *Locating cultural work: The politics and poetics of rural, regional and remote creativity*. London: Palgrave Macmillan.
———. 2018. Craft entrepreneurialism and sustainable scale: Resistance to and disavowal of the creative industries as champions of capitalist growth. *Cultural Trends* 27 (5): 313–326.
Mar, P., and K. Anderson. 2010. The creative assemblage: Theorizing contemporary forms of arts-based collaboration. *Journal of Cultural Economy* 3 (1): 35–51.
McLean, H. 2017. Hos in the garden: Staging and resisting neoliberal creativity. *Environment and Planning D: Society and Space* 35 (1): 38–56.
McRobbie, A. 2016. *Be creative*. London: Polity.
Mould, O. 2018. *Against creativity*. London: Verso.
Munro, E. 2016. Developing the rural creative economy 'from below': Exploring practices of market-building amongst creative entrepreneurs in rural and remote Scotland. *M/C Journal* 19 (3): 1071. http://journal.media-culture.org.au/index.php/mcjournal/article/view/1071. Accessed 3 Feb 2020.
———. 2017. Building soft skills in the creative economy: Creative intermediaries, business support and the 'soft skills gap. *Poetics* 64: 14–25.
Nail, T. 2017. What is an assemblage? *SubStance* 46 (1): 21–37.
Njaradi, D. 2014. From employment to projects: Work and life in contemporary dance world. *Text and Performance Quarterly* 34 (3): 251–266.
Nurmis, J. 2016. Visual climate change art 2005–2015: Discourse and practice. *Wiley Interdisciplinary Reviews: Climate Change* 7 (4): 501–516.
Oakley, K. 2014. Good work? Rethinking cultural entrepreneurship. In *Handbook of management and creativity*, ed. C. Bilton and S. Cummings, 145–159. Cheltenham: Edward Elgar.
Oakley, K., M. Ball, and M. Cunningham. 2018. *Everyday culture and the good life. CUSP working paper no 9*. Leeds: University of Leeds.
O'Brien, D. 2014. *Cultural policy: Management, value and modernity in the creative industries*. London: Routledge.
Roberts, E., and L. Townsend. 2016. The contribution of the creative economy to the resilience of rural communities: Exploring cultural and digital capital. *Sociologia Ruralis* 56 (2): 197–219.
Scharff, C. 2016. The psychic life of neoliberalism: Mapping the contours of entrepreneurial subjectivity. *Theory, Culture & Society* 33 (6): 107–122.
Schlesinger, P., M. Selfe, and E. Munro. 2015. *Curators of cultural enterprise: A critical analysis of a creative business intermediary*. London: Springer.
Siebert, S., and F. Wilson. 2013. All work and no pay: Consequences of unpaid work in the creative industries. *Work, Employment and Society* 27 (4): 711–721.
Taylor, S., and K. Littleton. 2016. *Contemporary identities of creativity and creative work*. London: Routledge.
Waitt, G., and C. Gibson. 2013. The Spiral Gallery: Non-market creativity and belonging in an Australian country town. *Journal of Rural Studies* 30: 75–85.

Correction to: Cultural Industries and the Environmental Crisis

Kate Oakley and Mark Banks

Correction to: K. Oakley, M. Banks (eds.), *Cultural Industries and the Environmental Crisis*, https://doi.org/10.1007/978-3-030-49384-4

Chapter 8

The original version of this chapter was inadvertently published non-open access. It has now been updated to open access under a CC BY 4.0 license and the copyright holder has been changed to "The Author(s)". The book has also been updated with this change.

Chapter 12

The original version of this book was inadvertently published without the last chapter, 'Exhaustion' as a Driver of Change in the Creative Economy, by Ealasaid Munro. This chapter has now been included in this version of the book.

Front matter

The original version of this book was inadvertently published without list of contributors in front matter. It has been included now with all the contributors in this title.

The updated online versions of these chapters can be found at
https://doi.org/10.1007/978-3-030-49384-4_8
https://doi.org/10.1007/978-3-030-49384-4_13
https://doi.org/10.1007/978-3-030-49384-4

K. Oakley, M. Banks (eds.), *Cultural Industries and the Environmental Crisis*,
https://doi.org/10.1007/978-3-030-49384-4_12

The manufacturer's authorised representative in the EU is Springer Nature Customer Service Centre GmbH, Europaplatz 3, 69115 Heidelberg, Germany. If you have any concerns regarding our products, please contact ProductSafety@springernature.com

Printed and bound by CPI Group (UK) Ltd, Croydon, CR0 4YY
28/09/2026
02224092-0011